John Fitzgerald
Peter Gorm Larsen
Paul Mukherjee
Nico Plat
Marcel Verhoef

Validated Designs for Object-oriented Systems

John Fitzgerald
Peter Gorm Larsen
Paul Mukherjee
Nico Plat
Marcel Verhoef

Validated Designs for Object-oriented Systems

With 89 Figures

 Springer

John Fitzgerald
Centre for Software Reliability, University of Newcastle upon Tyne, UK

Peter Gorm Larsen
Systematic Software Engineering A/S, Denmark

Paul Mukherjee
Systematic Software Engineering A/S, USA

Nico Plat
West Consulting B.V., The Netherlands

Marcel Verhoef
Chess Information Technology B.V. and Radbound University Nijmegen, The Netherlands

British Library Cataloguing in Publication Data
A catalogue record for this book is available from the British Library

Library of Congress Cataloging-in-Publication Data
Validated designs for object-oriented systems / John Fitzgerald ... [et al.].
    p. cm.
  Includes bibliographical references and index.
  ISBN 1-85233-881-4 (alk. paper)
    1. Object-oriented methods (Computer science)   2. Formal methods (Computer science)
  3. System design.   I. Fitzgerald, John, 1965–
  QA76.9.O35V35   2004
  005.1'17—dc22   2004056526

ISBN 1-85233-881-4
Springer is a part of Springer Science+Business Media
springeronline.com

© Springer-Verlag London Limited 2005
Printed in the United States of America

Cover design: Joseph Piliero
Production: Michael Koy
Typesetting: Author
Printing: Sheridan Books

34/3830/543210   Printer on acid-free paper   SPIN 11000327

Preface

Object-oriented design is common in computing systems development. There is a plethora of techniques and an abundance of texts on the subject. Why, then, a book on *validated* designs for object-oriented systems? Our experience with specification, modelling and analysis techniques over the last decade tells us that object-oriented design is much more than mere "boxes and arrows." Systems developers should be able to combine techniques and tools to achieve designs that are not merely the subject of heated argument but can be improved by careful, rigorous and machine-supported analysis. This book describes an approach to the design of object-oriented systems which combines the benefits of abstract modelling with the analytic power of formal methods to give designs that can be rigorously validated and assured with automated support.

System modelling has a valuable contribution to make in the development of computing systems. It encourages abstract, systematic and comprehensive consideration of system aspects, including the description of functionality, interactions with the environment and temporal behaviour. The rising level of interest in modelling is evidenced by the central role being taken by model-driven development approaches within the software engineering communities, such as the model-driven architecture work by the Object Management Group.

Formal methods can bring further tangible benefits to systems and software development, including objectivity and rigour of analysis, without compromising abstraction. However, these benefits can only be realised in practice if formal techniques are carefully targeted. In particular, formal techniques cannot be applied in isolation but must work alongside other analysis and design tools. The successful application of these techniques is regularly reported. This book stands in the tradition of "Lightweight" formal methods. Its aim is to equip readers with the ability to take advantage of formal modelling techniques without necessarily having to deploy the whole panoply of formal methods. We show how a formal notation (VDM++) can be used to complement and enhance object-oriented class models, with the engineer free to move between the class structure view and the functional view in VDM++. The approach shows how formal techniques can be used to take practitioners on from where they are, rather than requiring a radical shake-up of existing practice.

Although still relatively novel, this synthesis of object-oriented design and formal methods is a credible technology which has been applied in a wide variety of industrial projects. The style and content of the book are intended to bear this out. The elements of formal model-oriented specification, basic types, collections, abstract specification of functionality and system structuring are illustrated by examples derived from industrial experience. The material in the text is supported by the use of industry-strength tools. Many exercises are woven into the text, and more will be made available on the accompanying Web site (http://www.vdmbook.com).

Structure of the Text

Part I provides motivation for modelling systems and enhancing models with descriptions of functionality. The introduction discusses the role of modelling in the development process and tool support. The reader's first glimpse of VDM++ comes in Chapter 2 when a realistic model of a chemical plant alarm management system is presented. Chapter 3 sets up the tool support that the reader will use throughout the remainder of the book.

Part II shows how the modelling approach is realised in a specific language, VDM++. Chapters 4 and 5 introduce the basics of modelling data and functionality in an object-oriented design. Chapters 6 to 8 introduce the key abstractions for modelling collections and relationships. The technical content is introduced through examples based on real applications of the technology (robots maneuvering around obstacles and congestion monitoring in road transport).

Part III is in many respects the core of the book. It situates the modelling approach in the industrial context by means of three case studies concentrating on different aspects of the application of modelling technology. Chapter 9 shows the use of modelling abstractions in understanding the architecture and operation of the famous Enigma cipher machine. Importantly, it also shows how established testing approaches can be applied in the context of a formal object-oriented design. Chapter 10 uses a continuous train speed monitoring function as an example to illustrate the links between the levels of detail in a formal model in VDM++ and the associated design-level class diagrams in UML. Chapter 11 gives an in-depth discussion of a real trading management ("back-office") system, including a review of the pros and cons of the modelling approach.

Part IV moves forward from the production of a model to deal with more advanced topics in the development process. Chapter 12 introduces the facilities for handling concurrency in VDM++. Chapter 13 looks at the facilities for gaining confidence in the accuracy of a model, both by simulation and by systematic consistency checking, a partly automated process. Chapter 14 discusses the implementation of models in Java.

Using the Book

The book is aimed at software engineers with some prior experience in object-oriented design/programming, including intermediate or advanced students and researchers studying object-oriented design. Readers are assumed to have some experience of programming in an imperative and object-oriented language such as Java or C++. Some familiarity with the basic concepts of class, attribute, method and inheritance is assumed. However, no specialised knowledge of formal methods is expected.

Readers new to object-orientation, for example those wishing to study formal modelling in conjunction with design in UML and programming in Java, can still benefit from the book. Such readers may omit the later parts of Chapter 9, all of Chapters 10 and all of Part IV on a first reading, revisiting these when greater confidence has been gained in programming.

Background to the Book

This book is a product of a very long line of research dating to the development of the Vienna Definition Language (VDL) in IBM's Vienna Laboratory in the 1970s. Originally targeted at language definitions and compiler development VDL and its successor the Vienna Development Method (VDM) gradually found a wider range of applications. Seminal texts, notably those by Dines Bjørner and Cliff Jones [Jones80, Bjørner&82, Jones90], presented, in a formal context, many notions which are mainstream now, such as pre- and postcondition specification, data and operation refinement. VDM's specification language, VDM-SL, achieved ISO standardisation of both its syntax and formal semantics in 1996.

From the 1990s, two of us (John Fitzgerald and Peter Gorm Larsen) were engaged with the development of formal methods strong enough for application outside specialised areas. We both had experience at applying such methods in industry, and we both wanted to bring this experience to bear on a subject that was widely regarded as arcane and irrelevant to professional practice. The collaboration led to a text [Fitzgerald&98] which introduced modelling principles using a VDM-SL subset and which was driven by examples of industrial application and supported by the commercial tools developed by IFAD A/S. Although the VDM-SL work was successful in a range of applications, it had already become apparent that the link between plain formal modelling and object-oriented design could be bridged. Nico Plat, Paul Mukherjee and, later, Marcel Verhoef, had been applying the extended object-oriented version of VDM, called VDM++, initially developed through the European Commission's Afrodite project and then in a wide range of commercial and research projects and in the development of tool support, again at IFAD.

This text represents a synthesis of those strands of work. Its aim is to present an achievable improvement to existing practice through the use of formal design techniques in the context of mainstream object-oriented design, and to demonstrate this through realistic case studies inspired by industrial practice.

Accompanying Web Site

Supporting material for the book, including VDMTools, manuals and further exercises is available from a variety of sources, many over the Web. The book's supporting website http://www.vdmbook.com acts as a portal, providing links to all the relevant sites.

Acknowledgements

We are grateful to many wonderful colleagues whose work has made it possible to produce this book, and all those whose patience we have tried in the process. We apologise in advance to any we have inadvertently omitted from our list.

IFAD A/S of Odense in Denmark developed the first industrial-strength tools for VDM and VDM++. We are profoundly grateful to all our colleagues at IFAD over many years for the opportunity to explore the potential of this technology.

Our work is constantly inspired by the experiences of practitioners in system design and formal methods. We are grateful to the technology users who have provided inspiration for the case studies reported in the book. We would particularly like to thank the Japan Future Information Technology and Systems Co. Ltd., and in particular Mr. Shin Sahara, for kindly providing information on the TradeOne development experience.

We have had positive and professional support from Beverly Ford, Catherine Drury, Michael Koy, Frank McGuckin, their colleagues and reviewers at Springer-Verlag during the production cycle for the book. We gladly acknowledge the help of many colleagues who commented on drafts of the text: Bernhard Aichernig, Ian Bayley, Barrett Bryant, Ian Cottam, Dave Hastings, Neil Henderson, Jozef Hooman, Kevin Lano, Mikhail Lebedev, Richard Moore, Luis Neves, Erik Toubro Nielsen, Oliver Oppitz, Stephen Paynter, Alexander Petrenko, Daan Rijsenbrij, Shin Sahara, Kim Sunesen, John Turner and Georg Weissenbacher.

Our respective institutes and employers have been very supportive during the production of the book. John Fitzgerald is grateful to Transitive Ltd. and the School of Computing Science at the University of Newcastle upon Tyne; Peter Gorm Larsen and Paul Mukherjee thank Systematic Software Engineering A/S; Nico Plat thanks West Consulting B.V.; Marcel Verhoef thanks Chess Information Technology B.V.

At a personal level, we are deeply grateful to our families and friends for their perseverance since the genesis of this book several years ago, to John Hudson, Michel Overbeeke, Marcelle van Valkenburg, and in particular to Yvonne Mukherjee and Margit Sandvang Larsen for their great hospitality during those many writing "conferences" in Odense!

Newcastle upon Tyne, UK *John Fitzgerald*
Århus, Denmark *Peter Gorm Larsen*
Fort Worth, Texas, USA *Paul Mukherjee*
Rotterdam, The Netherlands *Nico Plat*
Dordrecht, The Netherlands *Marcel Verhoef*

September 2004

Contents

Part II Modelling Object-oriented Systems in VDM++

Part III Modelling in Practice: Three Case Studies

Part I

Models and Software Development

1

Introduction

This chapter marks out the subject area for the book: the use of complementary modelling techniques to assist in the design of object-oriented software. Its aim is to introduce the reader to the notion of modelling, the elements of the object-oriented paradigm on which the book builds and the structure of VDM++ models.

1.1 The Challenge of Software Development

Software development can be both highly rewarding and deeply frustrating. It is rewarding because so much can be achieved with computers running good software. It is frustrating because the demands placed on software seem to outstrip the ability to develop it to satisfactory standards within time and budgetary constraints. Many of the challenges faced by software developers stem from complexity in the requirements for products and in the implementation media. It has become reasonable, for example, to expect a flight control system for an aircraft to be implemented largely in software on a distributed avionic hardware system composed of replaceable components. Predicting the functional and temporal behaviour of the hardware and software together is crucial to successful implementation if major rework costs are to be avoided after testing.

For software developers today, it is important to recognise that the program alone is not the product. Requirements change rapidly during development and between releases of software. It is increasingly important to have some means of recording design, design rationale and software structure and of providing traceability between them. Having such a traceable design set makes it possible to assess the consequences of changing requirements and changes in the software's operating environment.

A crucial challenge for successful software development is, therefore, that of marshalling complex requirements and creating traceable and justifiable designs at an affordable cost. The technology described in this book aims to help engineers meet that challenge.

The idea of system modelling as a means of mastering complexity through abstraction and rigorous, objective analysis is introduced (Section 1.2) and placed in the context of software development processes in Section 1.3. A software system's architecture is influenced by a chain of factors going back to the objectives of the organisation in which it is to be placed, and this is discussed in Section 1.4. The functionality within a structured model is described in Section 1.5. The combination of models of structure and functionality is motivated in Section 1.6. Finally Section 1.7 gives an overview of the structure of the remainder of the book.

1.2 Mastering Complexity: The Role of Modelling

This book is primarily about modelling as a means of mastering complexity in software development. A model is defined as

> A simplified or idealized description or conception of a particular system, situation, or process, often in mathematical terms, that is put forward as a basis for theoretical or empirical understanding, or for calculations, predictions etc.
> *The Oxford English Dictionary*

Models are routinely developed in many branches of engineering as a means of understanding the requirements for a system and assessing the capabilities of a design. Such models range from physical constructions developed for stress or wind tunnel testing to purely mathematical models subjected to theoretical or numerical analysis. In the world of software development, models range from the graphical representations of data (e.g. entity-relationship modelling) and functionality (e.g. data flow modelling) through program-like representations of algorithms (e.g. pseudocode) to the formal mathematical representation of functional and temporal behaviour using logics.

Successful models allow accurate analysis and prediction at reasonable cost. This is achieved through a combination of abstraction and rigour. Recall from the definition that a model is a "simplified or idealized" description. *Abstraction* is the omission from a model of detail that is not relevant to the model's purpose. Good abstraction depends on having a clear common understanding of the model's purpose. Before pen goes to paper, the modeller must spend time deciding what the model is intended to achieve. *Rigour* is the extent to which the model can be analysed objectively. Achieving a high level of rigour depends on the notation used for the model's expression. At one extreme, an ad hoc graphical notation used for drawings on a whiteboard is easy to employ but rather ambiguous: Such drawings rarely stand up to detailed questioning. At the other end of the spectrum, mathematical notations have a very well-defined semantics and are capable of objective and largely automated analysis but can be inaccessible and expensive to deploy in all but the most critical projects. The term "formal" is applied to modelling techniques that have a mathematical semantics so precise as to leave little or no room for disagreement about the properties of a model. This book charts a course between these extremes. A rigorous notation with ample tool support (VDM++) forms the basis of the modelling approach advocated here. Most importantly, it can be used as an optional enhancement to a graphical but well-defined notation (class diagrams in the Unified Modelling Language (UML)) rather than as an isolated technology.

Many different notations are used in a software development project, each with its strengths and weaknesses and each describing complementary views over the same system. Multiview paradigms, such as Kruchten's "4+1" view model [Kruchten95], use a range of notations for specific perspectives such as the *logical view* of end-user functionality, the *process view* of performance, the *physical view* of system topology and the *development view* of requirement and work allocation. It should be possible to work on models reflecting complementary views of a system while maintaining consistency between them. The characteristics of a successful modelling environment, then, are not just abstraction and rigour in the modelling language, but an ability to use distinct, focused, models in a coordinated and consistent way.

1.3 The Place of Modelling in Software Development

Software development is the process of implementing, in a programming language, a system that satisfies the requirements of a client. Those requirements may not have been clearly described at all, but they are usually set out in a range of notations including natural language, graphics and mathematics. Requirements will generally have been written in terms familiar to people who are expert in the application domain. The implementation and its accompanying documentation must take full account of the implementation medium, target language and computing system, concepts that will often be unknown to the application domain experts involved in requirements definition. Software development involves bridging the gap between these two worlds.

The process of development involves a wide range of activities, from requirements gathering through to testing and maintenance. These activities are typically arranged into development processes. Software development processes range from highly concurrent forms, in which many activities take place at once, to highly sequential processes with an official signoff at the end of one activity before the next activity can commence.

Development activities typically involve an element of synthesis, for example, developing requirements statements, design documents, code and test plans. These are complemented by analytic activities such as inspection and testing intended to establish that the synthesised product meets some criterion: correctness with respect to requirements, for example. Errors committed in early design stages but not detected quickly can lead to expensive reengineering as the design steps between commission and detection have to be repeated. Modelling techniques, applied in early development stages, provide an opportunity to detect defects close to the point at which they are introduced, hence limiting rework costs.

This book focuses on modelling and analysis techniques in early development stages, specifically design. Design is the activity of constructing an outline of the implementation from a statement of system requirements. It is the first point at which the requirements come up against the constraints of the implementation medium. As such, design needs to take account of some of the characteristics of the implementation medium. In particular, the structuring mechanisms in the target language may be incorporated at this stage.

The abstraction capability that comes with the use of a special-purpose modelling language allows us to construct models of systems at development stages when the implementation details may not be known. For example, if a modelling language supports the use of abstract data structures such as unbounded sets, we can model collections as sets of records, rather than having to worry about indexing into linked lists, handling heap storage etc. This allows us to represent requirements without overloading the model with detail specific to the implementation environment.

The use of a rigorous modelling language permits the analysis of a design at an early development stage, helping to detect the defects that otherwise might not come to light until much later, with all the attendant reworking. This ability to *validate* the model gives the developers increased confidence in proceeding to more detailed design and implementation. The IEEE Standard for Software Verification and Validation describes validation [IEEE1012-1998] as "In design and development, validation concerns the process of examining a product to determine conformity with user needs." The validation process therefore involves assessing the model against requirements that are often informally expressed. Using a formal modelling notation affords using a wider range of validation techniques than might be the case for a notation with less clearly expressed semantics. It is worth briefly contrasting validation with *verification*, which, in the terms of the IEEE Standard "concerns the process of examining the result of a given activity to determine conformity with the stated requirement for that activity." Verification assesses conformance of a particular development product against its usually relatively precise specification.

Object-oriented languages are increasingly popular as an implementation medium. As a result, determining the object/class structure for the implementation and allocating capabilities to the component parts are central to successful design. In order for a modelling language to assist with this process, it must be capable of representing not only data and functionality, but also object-oriented structure.

This book is about the use of modelling to produce abstract, rigorous design models for object-oriented systems. The models produced can be analysed thoroughly to give confidence in their integrity and faithfulness to the informally expressed requirements. Finally, they can also serve as a basis for implementation in an object-oriented programming language such as Java or C++. The following sections refine these ideas.

1.4 Modelling Structure: Object-oriented Systems

This book assumes some experience of object-oriented design or programming. Readers without this background are referred to any general text on software engineering [Pressman04, Sommerville04]. There is wide variation in terminology and so the basic concepts in the object model underpinning VDM++ are reviewed briefly in this section.

In the object-oriented paradigm, systems are viewed as collections of interacting objects. An object is an entity that typically contains some data and provides services to other objects through an interface. The data associated with an object are called *attributes*: They represent the internal variables or state of the object. The services provided by the object are termed *operations* or *methods*. These constitute the object's interface to the outside world. The attributes and operations are *encapsulated* in the object: A reference to the object gets not just data but also the ability to call the operations. Access to the data within an object is normally only possible through the published operations, not directly.

Systems typically contain many objects with the same attributes, albeit with different values, and with the same operations. A group of objects with common features is termed a *class*. A class can be thought of as a template describing the characteristics (attributes, operations) shared by its member objects. The objects of a class are said to *inherit* the characteristics. A class may itself inherit characteristics from another class, termed its *superclass*. A class inheriting from a superclass is termed a *subclass*. This leads in some systems to a *class hierarchy* of inheritance.

As an illustration of these concepts, consider a simple example. An airline wishes to build a computer system to track the maintenance of aircraft. The airline has a collection of aircraft, some of which are in maintenance and therefore out of service. Each aircraft out of service has a maintenance schedule, indicating what tasks are to be performed. For example, there may be several painting jobs to be performed, some electrical checks to be done, some engine maintenance. Typically, these tasks will have to be done in some particular order. The airline has teams of engineers who perform the jobs, with a team attached to an aircraft at each time. Each task in the aircraft's maintenance schedule will have some engineers assigned to it. The class structure for this example is presented in the next section, introducing the graphical notation for class structures that are used in subsequent chapters.

Class Diagrams

The class diagrams used in this book are based on those supported in UML [Booch&99, Fowler99]. A class diagram provides a graphical overview of the static structure of a collection of selected classes and their internal relationships.

In a class diagram, a class is represented as a box divided into three compartments showing, from top to bottom, the name of the class, its attributes and its operations. Figure 1.1 shows the representation of a class of aircraft. Each object of the class represents an aircraft and contains attributes representing the aircraft model, serial number and number of miles flown. Each attribute has a name, to the left of the colon, and a data type, to the right of the colon. One operation is available for an aircraft object: to update the number of miles flown. The operation takes an input but does not return a result; it just updates the miles flown attribute.

Suppose that a class is created to represent aircraft in maintenance. These are still aircraft, and they inherit aircraft attributes, but they also have start and end dates for the maintenance period as additional attributes, plus an operation to update the end date if tasks are running behind or ahead of schedule. The inheritance relationship is indicated by a closed arrow, as shown in Figure 1.2.

Other classes not in an inheritance relationship may appear on a class diagram. For example, classes representing maintenance teams, engineers and tasks could be added

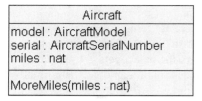

Fig. 1.1: Aircraft class

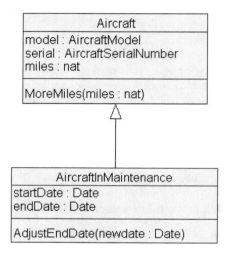

Fig. 1.2: Aircraft classes in an inheritance relationship

as shown in Figure 1.3. Noninheritance relationships between classes are shown as *associations*, indicated by arrows with open heads as shown in Figure 1.3. The relationships are labelled with meaningful names and should be read in the direction of the arrow. For example, the association at the bottom of Figure 1.3 shows that an engineer "works on" an aircraft. The numeric labels give the *multiplicity* of the association, indicating lower and upper bounds on the number of objects at the destination end of the association that can be related to objects in the class at the source end. Further qualifications can be applied to associations. These are indicated in braces beside the association. The `ordered` constraint, illustrated on the "tasks" association in Figure 1.3, indicates that objects at the destination end have some ordering, in this case ordering among the scheduled tasks.

Associations can be further structured by *qualifications*. A refinement of part of the class diagram is shown in Figure 1.4. The "tasks" association is qualified by the data type "TaskType" to indicate this when the association is indexed with different types of tasks. Thus, in following the association from source to destination, it will be

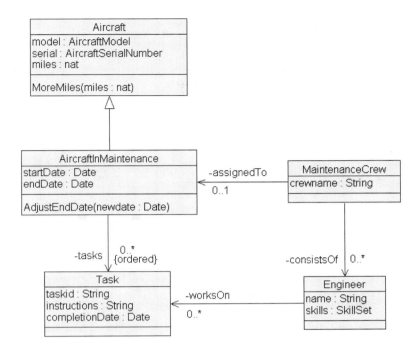

Fig. 1.3: Aircraft maintenance system classes

possible to find an ordered collection of tasks for each task type (e.g., repainting tasks, electrical tasks).

Given the importance of encapsulation in object-oriented design, it is worth considering the accessibility of each object's data and operations to objects of other classes. Figure 1.5 shows the aircraft maintenance system extract decorated with *access modifiers*. In the models described in this book, there are three levels of visibility for attributes, operations and associations. These will be further described in Section 4.2.1.

The class diagram gives an indication of the proposed structure of an object-oriented system implementing requirements. The presentation here has been restricted to basic features: classes with their attributes and operations, inheritance relationships and associations. Further features will be added to the notation as they are required in subsequent chapters.

The class diagram is a widely used tool. Although it outlines the structure of an object-oriented design, analysis of this structural model is limited to considering just the structure itself. What about the functionality to be supplied? What exactly should the operations do? When should they work and what constraints should they respect? In order to gain the ability to model these aspects, the class diagram is supplemented

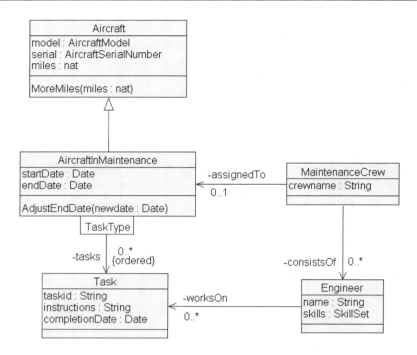

Fig. 1.4: Aircraft maintenance system class diagram with qualified association

with a model that gives details of the data and operations associated with each class. This enhanced model is expressed in VDM++.

1.5 Modelling Data and Functionality: VDM++

The origins of VDM++ lie in VDM, the Vienna Development Method, which grew out of work conducted at the IBM Development Laboratories in Vienna in the mid-1970s [Jones99]. At the core of VDM is a formal modelling language, the VDM Specification Language (VDM-SL), the syntax and semantics of which were standardised by ISO in 1996 [ISOVDM96]. The formal proof theory underpinning the bulk of the language is set out in [Bicarregui&94]. An introduction to modelling in the language is given in Fitzgerald and Larsen's 1998 text [Fitzgerald&98]. The Afrodite project (1992-4) developed extensions to VDM-SL to accommodate object-oriented structuring and the handling of concurrency. In more recent years, there has been considerable work on improving the quality of tool support for VDM++ and gaining experience with a wider range of industrial applications. The remainder of this section gives a brief overview of the elements of a VDM++ model. The issues raised here will be dealt with in depth in later chapters.

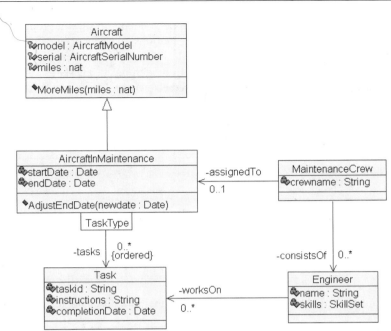

Fig. 1.5: Aircraft maintenance system class diagram with access modifiers

The Structure of a VDM++ Model

Class definitions are the building blocks of models in VDM++. Objects are *instances* of classes, each object having its own identity and each supplying the same kind of functionality to the outside world. A class definition is bounded by delimiters giving the class name, and between which the members are defined, as in the following example:

```
class Aircraft

  member declarations

end Aircraft
```

A VDM++ model is structured as a *collection of classes* that may be linked by inheritance and association relationships. The superclasses of a given class are introduced using the is subclass of keyword, e.g.,

```
class AircraftInMaintenance is subclass of Aircraft
```

The *member declarations* inside a class can be divided into several groups: definitions of types, values (constants), functions, operations and concurrency constraints. Each group is considered briefly here.

Type Definitions: Data types are central to modelling in VDM++. Type definitions associate names with significant types used elsewhere. For example, it would be possible to define a type `AircraftModel` to represent the kinds of aircraft in the example maintenance system. The type could be represented as an enumeration of the possibilities:

```
class Aircraft

types
AircraftModel = <BOEING767> | <BOEING747_400> |
                <AIRBUSA310> | <AIRBUSA320>

end Aircraft
```

The serial number might be represented by a type that models character strings, e.g.,

```
class Aircraft

types
AircraftModel = <BOEING767> | <BOEING747_400> |
                <AIRBUSA310> | <AIRBUSA320>;

SerialNumber = seq of char
...
end Aircraft
```

where the ellipsis ". . ." indicates that more definitions may be present.

VDM++ allows the designer to record restrictions on data values that might otherwise remain unstated. For example, there may be a restriction that serial numbers are nonempty and at most 20 characters long. Such a property that always applies to all elements of a type is known as a *data type invariant*, or just *invariant* and can be stated as a Boolean expression (i.e., an expression that yields `true` or `false`):

```
class Aircraft

types
AircraftModel = <BOEING767> | <BOEING747_400> |
                <AIRBUSA310> | <AIRBUSA320>;
```

```
AircraftSerialNumber = seq of char
inv snum == len snum > 0 and len snum <= 20;
...
end Aircraft
```

Instance Variables: These describe the attributes that are present in every object instantiated from the class. Each instance variable has an associated type, e.g.,

```
class Aircraft
...

instance variables

   model  : AircraftModel;
   serial : AircraftSerialNumber;
   miles  : nat;

...
end Aircraft
```

The instance variables may be restricted by an invariant, typically recording consistency constraints that link the variables. For example, there could be a restriction stating that all Airbus aircraft have serial numbers starting with the character 'A'. This can be stated using basic logical operators, including "or" and "=>", the logical implication operator (logic operators are introduced in Section 4.4.1):

```
class Aircraft

instance variables

private model  : AircraftModel;
private serial : AircraftSerialNumber;
private miles  : nat;
inv (model = <AIRBUSA310> or model = <AIRBUSA320>) =>
    serial(1) = 'A';

...
end Aircraft
```

Value Definitions: These are constants declared to be used in the VDM++ model.

Function Definitions: Functions describe parts of the behaviour of objects belonging to the class. A function takes parameters and in return delivers a result in accordance with the function's definition.

In VDM++ functions can be defined either *implicitly* or *explicitly*. Implicit function definitions simply state the properties of *what* the function should do without giving a specific way of computing it. This is done by precisely stating the properties relating the parameters of the function and the result. Explicit function definitions, on the other hand, use an algorithmic description expressing *how* to calculate the result for the given parameters. It is important to note that function definitions in VDM++ cannot make use of any instance variables. The functions themselves are pure and have no side effects. The body of an explicit function consists of an expression. The result of a function call is the body expression evaluated with the actual parameter substituted for the formal parameters.

Operation Definitions: Operations, like functions, describe some part of the behaviour of objects of the class. An operation also takes input parameters and can deliver a result. In addition, it may update the instance variables in the object or another object referred to from within the present object.

In VDM++ operations can also be defined either *implicitly* or *explicitly*. An implicit operation definition characterises the operation by giving the properties of the result in relation to the inputs, including the changes to the instance variables. In an explicit operation definition, an algorithmic body describes the computation performed to the inputs and on the instance variables. The body of such an operation definition is termed a *statement* and bears a close resemblance to the kind of statement seen in imperative programs in languages such as Java, in which programs may have side effects, changing the instance variables during their execution.

Thread Definitions and Synchronisation Constraints: VDM++ permits the modelling of concurrent or parallel computing systems. Thread definitions and synchronisation constraints describe the sequences of parallel computations that can be performed and define any synchronisations between them. These features are discussed in depth in Chapter 12.

The member declarations inside a class are organised into a sequence of blocks, each preceded by a keyword, e.g. `functions`. The ordering of the definitions is entirely up to the user: There is no requirement to define constructs before they are used. The only exception is for value definitions, which must be defined before use.

1.6 Using VDM++ in Object-oriented Design

Although based on formal methods, modelling in VDM++ is not a mathematical exercise divorced from the real world of software development. Considerable experience has been accumulated in the use of modelling techniques including VDM++ and VDM-SL, on industrial projects. So where is this technology useful in practice? Although software modelling has mainly been advocated for requirements analysis, it is becoming apparent that models can be beneficial in design and testing.

During design phases of development, the use of a modelling language such as VDM++ permits the precise description of desired functionality. The ability to formalise properties as invariants on instance variables and pre- and postconditions on operations means that these important constraints are not forgotten in subsequent development. Rigour in the semantics of VDM++ makes it possible to check internal consistency of models using type checking, e.g., checking signatures of operations and types in declarations (attributes and associations in UML), and other forms of consistency analysis that are susceptible to automated support (see Chapter 13).

Models can serve as "proving grounds" for trying alternative solutions (algorithms, data structures, etc.) before a commitment is made to production quality code, especially if the models are executable. Rapid prototyping techniques allow the model to be executed together with a graphical front end or existing software for which a new component is being specified.

Although models usually emphasise *what* is to be computed rather than *how*, they can provide guidance in design and coding. The ability to record restrictions and consistency constraints in a model means that such conditions can form part of the consistency checking in the application software itself, in assertions compiled into preproduction versions of code for example. During defect detection, a model can serve as either a source of possible tests or an oracle against which test outcomes can be assessed (see Chapters 9 and 13). The modelling techniques used in this book are based on VDM++, and they have been applied effectively in all of these areas.

1.7 Book Organisation and Content

This book describes a design approach in which class diagrams are used alongside appropriate data and functional abstractions in VDM++. The aim here is to provide tools and techniques that can be used to augment and improve existing practice. The approach is quite general and is by no means limited to VDM++. However, the extent of experience and the strength of tool support for VDM++ make it an appropriate modelling language to use in this way. It should also be stressed that the book deals primarily with design rather than requirements analysis activities.

The text is divided into four parts. The remainder of Part I introduces the combination of formal textual models of functionality with graphical object-oriented design using UML. This is done through a substantial example (Chapter 2), serving as a "guided tour" through VDM++ and, in Chapter 3, providing an introduction to the tools supporting the book that allow readers to construct their own models. Part II (Chapters 4 to 8) shows how the approach is realised in VDM++, dealing in turn with each of the major abstractions for handling different kinds of collection and association. Each chapter in Part II uses an example based on a real industrial application of the VDM++ technology, albeit in a simplified form. Part III (Chapters 9 to 11) contains three case studies, again based on industrial practice, showing the technology in practice. This part culminates in a report of the industrial use of VDM++ to develop TradeOne, a back-office system for trading in financial markets. The TradeOne story illustrates many of the significant characteristics of applying VDM++ in practice to assist in the design of larger-scale projects. Part IV moves the story on from models to code, considering the validation of models once they have been developed, ways of dealing with concurrency, and the implementation of models in Java.

Finally the book ends with a collection of solutions to exercises, a bibliography, a list of acronyms used and two indices: one for subjects and one for VDM++ definitions presented in the book.

1.7.1 Conventions Used in This Book

Several text and layout styles are used for the various kinds of information presented. VDM++ models are presented as follows:

```
class class name

  different kinds of definitions

end class name
```

The VDM++ text is shown in a typewriter font with keywords in boldface. The *emphasised* parts are used as meta text to introduce new VDM++ concepts. This convention will also be used for VDM++ extracts without class delimiters. Annotations are sometimes used within the model and these are preceded by the double hyphen character "--".

Java code is presented as shown in the following example:

```
public A () throws CGException {
  try {
    i = new Integer(0);
  }
  catch (Exception e){
    e.printStackTrace(System.out);
    System.out.println(e.getMessage());
  }
}
```

The typewriter font is used again but single frames are used for the box. VDM++ and Java code are always presented in boxes with rounded corners. Commands to tool prompts are shown in boxes with sharp corners. For example, the following is an example of a command to the VDMTools command line prompt:

```
vppde options and parameters
```

Whenever references are made to specific tool features, a sans serif font is used (e.g., interpreter).

Some chapters include special guidelines or rules, for example, for describing the translations between class diagrams and VDM++ models. Such material is placed in single-framed boxes with sharp corners. For example:

> **Guideline 1:** Try to make explicit operation definitions in VDM++ precise and clear, and yet abstract compared to code written in a programming language.

Most chapters include exercises. These are all numbered relative to their chapter. Solutions for most of the exercises are given at the back of the book whereas the solutions that involve large models are left to the book's Web pages. More challenging exercises are marked with a star ($\star$). Except for tool familiarisation exercises, these should be attempted by hand before using the tool support. As with any formalism, it is vital to master it in one's own mind before placing confidence in tools.

The subject index at the end of the book gives, in bold type, a pointer to the definition or introduction of each concept. The other page numbers refer to additional information on the topic.

1.7.2 Overview of Examples

The book uses many different examples drawn from a variety of application areas. The larger examples have been designed to illustrate features of the modelling technology that arise in practice, as well as language-specific points. In some cases, the text presents abstracts from larger models published on the book's supporting Web site. To give a flavour of the diversity of examples used, the major ones are listed here:

Chemical Plant: Chapters 2 and 3 use an example based on a simple system for calling experts out to deal with alarms produced by a chemical plant. The example shows many of the features of VDM++ in miniature and provides a "guided tour" of VDM++ and its tool support.

Electronic Patient Records: Chapters 4 and 5 present the basics of the modelling language using a model of parts of an electronic health record system.

Robot Controller: Chapter 6 uses a model of a robot controller to show how unordered collections are used, in this case to model motion through a space.

Congestion Warning System: Chapters 7 and 8 illustrate the use of ordered collections and relationships by building progressively more sophisticated models of a congestion monitoring system for traffic.

Enigma: Chapter 9 is the first of the substantial case studies showing the modelling technology in practice. The example, from the field of cryptography, is that of the famous Enigma cipher which played a significant role in the Second World War.

Control Speed Limitation and Monitoring: Chapter 10 concentrates on the link between UML and VDM++, and model consistency. It uses an example from the railway domain – a system for monitoring and controlling train speed in critical areas.

TradeOne: Chapter 11 presents some results from the application of the modelling technology to the development of a back-office solution developed in Japan for trading in stock options. The results include data on development costs and product qualities.

POP3 Server: Chapters 12 to 14 introduce some more advanced aspects of modelling based on the standard POP3 protocol enabling clients to fetch email from a server.

1.8 Summary

The aim of this book is to show how object-oriented design can be enhanced by developing models that are abstract, precise and susceptible to analysis. Object-oriented systems are structured into classes (with attributes and operations), inheritance and association relations. Such structures can be conveniently represented by class diagrams. However, the ability to model the data structures and functional behaviour of such systems in VDM++ is an asset to design.

2

Building a Model in VDM++: An Overview

The aim of this chapter is to give a first sight of the activities involved in producing an enhanced system model using VDM++ and UML class diagrams. This is done by way of an extended example based on an alarm management system. Although many features of VDM++ are introduced at a superficial level in this chapter, they are described in greater detail in subsequent parts of the book.

2.1 Constructing a Model

This chapter provides an overview of the activities involved in deriving, validating and implementing models. In particular, it provides initial guidance on how to construct models using the combination of VDM++ and UML class diagrams. Although this is a great deal of material for a single chapter, there is no need to understand all the details of the VDM++ features demonstrated, as these will be covered in depth later in the book.

Any modelling task must begin with a thorough consideration of the purpose for which the model is being constructed. This guides the often tricky choices to be made between alternative abstractions. For example, when designing an automobile, the purpose of one model might be to predict the aerodynamic properties of the vehicle, the purpose of another to determine the effects of different brake systems, the purpose of a third to choose the most effective layout for instruments, and so on. Each of these models will abstract away from different aspects of the car that might be relevant to other models. For example, the aerodynamics model would not likely include details of the operation of the brake pedal. The abstraction determines which details will be represented and which will be ignored because they are not relevant to the analysis.

It can be difficult to know how to begin constructing a useful model. Developers rarely have a complete set of requirements from which to start and may use techniques for deriving use cases, class diagrams and sequence diagrams to help drive these out. The approach we advocate here does not supplant that activity. The following steps are intended as a guide to developing a formal model; they should certainly not be seen as a full-fledged method:

1. Determine the purpose of the model.
2. Read the requirements.
3. Analyse the functional behaviour from the requirements.
4. Extract a list of possible classes or data types (often from nouns) and operations (often from actions). Create a dictionary by giving explanations to items in the list.
5. Sketch out representations for the classes using UML class diagrams. This includes the attributes and the associations between classes. Transfer this model to VDM++ and check its internal consistency.
6. Sketch out signatures for the operations. Again, check the model's consistency in VDM++.
7. Complete the class (and data type) definitions by determining potential invariant properties from the requirements and formalising them.
8. Complete the operation definitions by determining pre- and postconditions and operation bodies, modifying the type definitions if necessary.
9. Validate the specification using systematic testing and rapid prototyping.
10. Implement the model using automatic code generation or manual coding.

As indicated in Chapter 1, this book focuses on design rather than requirements analysis. Different methodologies have different ways of capturing the requirements for a system. For example, this can be dealt with using traditional use cases from UML [Cockburn00, Rosenberg&99] or by writing some VDM definitions without object-oriented structuring [Fitzgerald&98].

2.2 A Chemical Plant Example

This section presents the requirements for a simple alarm system for a chemical plant. It forms a running example that serves to illustrate the process described earlier and to introduce elements of the VDM++ modelling language. Although the modelling process is described here as though it were a single-pass activity, a real development would usually be iterative. Furthermore, because the example used for our illustration here is small, step 3 from the list will be omitted entirely.

The example is inspired by a subcomponent of a large alarm system developed by IFAD A/S and introduced in [Fitzgerald&98]. A model of the system will be developed and validated using the facilities of Rational Rose® and VDMTools®, including the Rose-VDM++ Link, enabling a graphical overview of the model in the form of UML class diagrams. Chapter 3 provides an interactive and hands-on tour of the tools available for supporting the development of the model.

Imagine that you are developing a system that manages the calling out of experts to deal with operational faults discovered in a chemical plant. The plant is equipped with sensors that are able to raise alarms in response to conditions in the plant. When an alarm is raised, an expert must be called to the scene. Experts have different qualifications for coping with different kinds of alarms. It has been decided to produce a model to ensure that the rules concerning the duty schedule and the calling out of experts are correctly understood and implemented. The individual requirements are labelled R1, R8 for further reference:

R1. A computer-based system is to be developed to manage the alarms of this plant.

R2. Four kinds of qualifications are needed to cope with the alarms: electrical, mechanical, biological, and chemical.

R3. There must be experts on duty during all periods allocated in the system.

R4. Each expert can have a list of qualifications.

R5. Each alarm reported to the system has a qualification associated with it along with a description of the alarm that can be understood by the expert.

R6. Whenever an alarm is received by the system an expert with the right qualification should be found so that he or she can be paged.

R7. The experts should be able to use the system database to check when they will be on duty.

R8. It must be possible to assess the number of experts on duty.

In the next section the development of a model of an alarm system to meet these requirements is initiated. The purpose of the model is to clarify the rules governing the duty roster and calling out of experts to deal with alarms.

2.3 Choosing a Structure Using UML Class Diagrams

This section shows how a first model of the alarm system could be developed. A UML class diagram is constructed from the informal requirements and VDMTools are used to check its consistency. In this and subsequent activities, methodological guidelines worth stressing will be highlighted by a small boxed statement.

2.3.1 Creating a Dictionary

Following the first steps of the outline process, a list of candidate classes and data types could be extracted from the requirements description. The designer's notes on reading the requirements might look like this:

Potential Classes and Types (Nouns)

- *Alarm: required qualification and description*
- *Plant: the entire system*
- *Qualification (electrical, mechanical, biological, chemical)*
- *Expert: list of qualifications*
- *Period (whatever shift system is used here)*
- *System and system database? This is probably a kind of schedule.*

Potential Operations (Actions)

- *Expert to page: when an alarm appears (what's involved? alarm operator and system)*
- *Expert is on duty: check when on duty (what's involved? expert and system)*

- *Number of experts on duty: presumably given period (what's involved? operator and system)*

These constitute a kind of elementary dictionary of the main terms used in the requirements. It is often advisable to create a dictionary before starting to sketch models, in order to identify similarities and redundancies before beginning to use tool support. Although the preceding list will suffice for a small example, a dictionary in a real development would be much more detailed. The potential classes and types identified in the dictionary could then form the basis of a class diagram, whereas the potential operations might be described as use cases. The reader is referred to standard texts on UML for a full discussion of use cases.

2.3.2 Sketching Class Representations

To refine the dictionary definitions further, it is necessary to explain what the difference is between a class and a type. In pure object-oriented approaches there is no distinction between the two, with the result that one often gets a very large and complex collection of classes. VDM++ makes a distinction between classes and types, and this is used to reduce the number of entities in the model. Types are used when the only interest lies with the data whereas classes are used when data and their associated functionality cannot be separated.

> **Guideline 1:** Nouns from a dictionary should be modelled as types if, for the purposes of the model, they need have only trivial functionality in addition to read/write.

The term "qualification" from the chemical plant dictionary is an example of a noun that is best modelled as a type. Qualification is an enumeration type because it holds just four possible values and has only limited functionality. All that is required of a qualification is that it should be possible to check its equality with another qualification. If the scope of the model included the different kinds of actions different experts could carry out depending on their qualification, it would have been more natural to model qualification as a class with four subclasses.

Now "alarm" and "expert" from the dictionary should be considered. Both of these correspond to groupings of real objects in the physical world of this example, which coexist and act independently. Therefore these will both be modelled as classes:

Alarm is a class with attributes *required qualification* and *description*. Descriptions are strings that can be considered a type for the same reasons that the concept of Qualification could be considered as a type (see Figure 2.1).

Expert is a class with *qualifications* as an attribute. The requirements state that there is a list of qualifications. The word "list" implies some kind of ordering but the requirements do not indicate how this ordering should be arranged. In a real development this would have to be clarified with the "customer", but for the time being it will be assumed that there is no specific ordering which has importance for the desired functionality. In fact, the word "list" is often used to indicate collections in which no particular ordering is required. In VDM++ it is possible to define an unordered collection using a special 'set of' type constructor. Much more information about such sets will be given in Chapter 6. The Alarm and Expert classes are shown in Figure 2.1.

Fig. 2.1: The Alarm and Expert classes

It is necessary to create some association between alarms and the schedule of experts. To model such associations it is advisable to create an overall system class: a main class, containing associations to the other classes.

Guideline 2: Create an overall class to represent the entire system so that the precise relationships between the different classes and their associations can be expressed there.

In this example the main class will be called Plant. Recall that the purpose of this model is to clarify the rules concerning the duty schedule and the calling out of experts to deal with alarms. Two aspects of the "plant" are important in this respect: the *schedule* of experts on duty and the collection of (registered) possible *alarms*. Other aspects of the plant are not included in the abstraction. Consider the alarms first. It is necessary to have an association from the Plant class to the Alarm class and, because more than one alarm will be present in this system, it is necessary to use a multiplicity with this association. The association is called alarms (see Figure 2.2).

Guideline 3: Whenever an association is introduced consider its multiplicity and give it a rôle name in the direction in which the association is to be used.

Consider the schedule that must hold a number of experts allocated to periods. For each period, zero or more experts can be on duty. Thus it is necessary to use an association qualified with period that has a multiplicity of zero or more. This association is called schedule.

Guideline 4: If an association depends on some value, a qualifier should be introduced for the association. The name of the qualifier must be a VDM++ type.

Nothing is mentioned about periods in the requirements, but it is not necessary to know much, given the focus of the model. Periods just represent units of time to which experts are allocated: The model does not constrain the choice of periods or shift patterns. The period is modelled as a type rather than a class because it does not encapsulate any significant functionality.

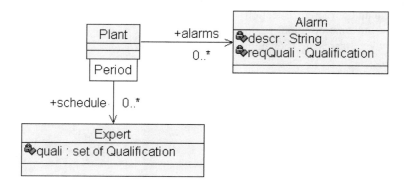

Fig. 2.2: Initial class diagram

A class diagram for the three classes defined so far is shown in Figure 2.2.

The diagram was drawn using Rational Rose, adding attributes and establishing relationships between the classes. The attributes and their types are entered via the templates provided by Rose. The types are VDM++ types, e.g., the set type constructor "set of" and class names and other type identifiers. All of this could equally well have been done directly in VDM++, and sometimes developers find it faster to write the textual model first and produce the graphical visualization using the Rose-VDM++ Link. The detailed mapping rules between UML class diagrams and VDM++ are explained in Chapter 10.

It is important to remember that there is no such thing as "the right model." For example, if it was desired to clarify the business processes concerning the use of this alarm subsystem, perhaps as part of a larger computer system, then the model would probably be different. For example, it might be necessary to focus more on how schedules are constructed and maintained or on the reactive way in which the alarms are reported from different sensors.

Now that a UML class diagram is available, it is possible to map it automatically to a VDM++ model skeleton, using the Rose-VDM++ Link. The resulting VDM++ outline class definitions are discussed next.

The Class Plant

```
--
-- THIS FILE IS AUTOMATICALLY GENERATED!!
--
-- Generated at 27-jul-04 by the Rose VDM++ Link
--
class Plant

instance variables

public alarms    : set of Alarm;
public schedule : map Period to set of Expert;

end Plant
```

This outline class definition has the typical VDM++ class delimiters, as seen in Chapter 1. Within the class are two definitions of instance variables derived from the associations in the class diagram. The `alarms` instance variable uses the VDM++ set type constructor to represent the unordered "zero-or-many" association of alarms to a plant. Sets are introduced in detail in Chapter 6. The other instance variable, `schedule`, represents a qualified association as a VDM++ mapping. A mapping is like a table in which it is possible to look up with a key (in the domain) to get the associated value (in the range). Mappings are presented in more depth in Chapter 8. Earlier, the key is a period and the value is a set of experts. Note the `public` keyword before the two instance variables. This has been included because all associations in Rose are by default given public visibility, indicated by the small "+" symbol next to the rôle names for the two associations. To provide encapsulation, it is normally advisable to avoid granting public access to instance variables.

> **Guideline 5:** Declare instance variables to be `private` or `protected` to keep encapsulation. If nothing is specified by the user, `private` is assumed automatically.

Following Guideline 5 the two occurrences of `public` can simply be removed because the default in VDM++ is `private`. This will ensure the encapsulation of members of the class. The rules for this are explained further in Chapter 4. The comments at the beginning of the VDM++ extract are automatically inserted by the Rose-VDM++ Link. For brevity this part will be omitted in all subsequent examples in this book.

The Class Expert

```
class Expert

instance variables

private quali: set of Qualification;

end Expert
```

Attributes of classes in class diagrams become instance variables of VDM++ classes, just like associations from class diagrams. Note again that the VDM++ set type constructor is used to represent the association with a "zero-or-many" multiplicity for the quali instance variable. The private keyword appears because Rose by default treats attributes with private visibility.

The Class Alarm

```
class Alarm

instance variables

private descr   : String;
private reqQuali: Qualification;

end Alarm
```

Again, the Alarm attributes appear as instance variables in the VDM++ model.

Type Checking the Classes

> **Guideline 6:** Use VDMTools to check internal consistency as soon as class skeletons have been completed and before any functionality has been introduced.

Even though the VDM++ generated so far is just a skeleton, with no functionality represented in it, it is perfectly valid VDM++ and can be submitted to tools for elementary internal consistency checking, e.g., that all identifiers are well-defined. The three preceeding classes are not type correct, as there are three identifiers that are not defined: Period, Qualification and String (which is not a native VDM++ data type). Therefore the following three type definitions are inserted into the three VDM++ classes respectively.

```
class Plant
...
types

public Period = token;
end Plant
```

The type token is used to represent unspecified values. It contains an infinite collection of distinguishable elements. The only operator on tokens is the check of equality. A token type is used for Period because the exact representation for periods will be chosen in the final implementation and it is not considered significant in the current model. The type Period needs to be declared public because it must be available outside the Plant class.

Guideline 7: Tokens are useful for abstract models where unspecified values are to be used.

```
class Expert
...
types

public Qualification = <Mech> | <Chem> | <Bio> | <Elec>;
end Expert
```

Qualification is an enumeration type. In VDM++, this is technically represented as a union (indicated by the vertical bars) of quote types (the identifiers in angle brackets). A quote type contains just one value, which has the same name as the type itself. Both union and quote types are introduced in Chapter 4. In this model Qualification is defined inside the Expert class. This could also be done inside the Alarm class because an alarm requires a certain qualification to be dealt with.

```
class Alarm
...
types

public String = seq of char;

instance variables

descr    : String;
reqQuali : Expert'Qualification;
end Alarm
```

Note that the type of the instance variable reqQuali has been prefixed with the class name Expert because Qualification is defined there. This is called "name qualification" and is explained further in Chapter 4. Using the Rose-VDM++ Link, this change is updated at the Rose UML class diagram level automatically, as shown in Figure 2.3.

Fig. 2.3: The updated Alarm class

The type definitions do not have a counterpart in the UML class diagram and are therefore not translated. However, they are kept in the VDM++ model files and will not be deleted (or changed) by updates in the UML class diagram(s).

2.3.3 Sketching Signatures for Operations

The development is continued by adding operation signatures (the formal parameters and the result) at the class diagram level. All three operations listed in the directory belong naturally in the class Plant, because they are dependent on the schedule allocated there. The updated class diagram for Plant is shown in Figure 2.4 (note that the Expert class is left unchanged).

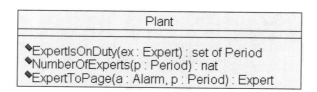

Fig. 2.4: The updated Plant class

The signatures of the operations are straightforward. For example, an operation such as ExpertToPage takes an alarm and a period as input and returns an expert as a result.

Guideline 8: Think carefully about the parameter types and the result type as this often helps to identify missing connections in the class diagram.

The updated skeleton VDM++ class is produced automatically.

```
class Plant

types

public Period = token;

instance variables

alarms   : set of Alarm;
schedule: map Period to set of Expert;

operations

public ExpertToPage: Alarm * Period ==> Expert
ExpertToPage(a, p) ==
  is not yet specified;

public ExpertIsOnDuty: Expert ==> set of Period
ExpertIsOnDuty(ex) ==
  is not yet specified;

public NumberOfExperts: Period ==> nat
NumberOfExperts(p) ==
  is not yet specified;

end Plant
```

The structure of the model developed so far is shown in Figure 2.5. It is possible to perform automated syntax and type checking on the model developed so far. However, far too much detail is omitted for the model to be of much analytic value. The next step is to use VDM++ to make the model more comprehensive.

2.4 Making the Model More Comprehensive

At this stage it is a good idea to review the model to check coverage of the requirements. Clearly the operations have not yet been considered in detail, so **R6-R8** are not fully covered. Otherwise the requirements seem to be covered reasonably well, except that requirement **R3** has not yet been documented anywhere:

R3 There must be experts on duty during all periods allocated in the system.

Requirement **R3** is certainly relevant to the model's purpose. It is important to know when experts are available in order to handle their call-out. So this aspect should

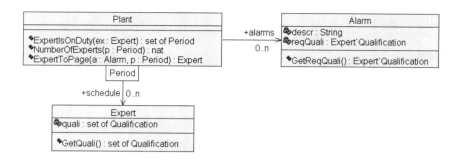

Fig. 2.5: Updated UML class diagram

certainly be included in the model. However, the limited apparatus introduced so far does not provide a mechanism for recording a constraint like this, certainly not at the level where it can be analysed systematically. The requirement can, however, be formulated precisely using VDM++.

2.4.1 Adding Invariant Properties

A period is said to be *allocated* in the system when it is present in the schedule for experts on duty. This can be formalised by saying that it is in the domain of the instance variable `schedule` already present in the model. In VDM++, the domain of the schedule is represented using an operator called `dom`. One can assert that a period p is allocated if

```
p in set dom schedule
```

To assert that for every allocated period p there should be an expert on duty, a VDM++ expression of the following form is used:

```
forall p in set dom schedule &
  there are experts on duty in p
```

This is an outline of a *quantified* expression, which states a property that must hold *for all* elements in the domain of the mapping. This kind of expression will be explained in more detail in Chapter 5. Next, what precisely does it mean when we say that there are experts on duty for a period? In terms of the model, it means that the range value associated with the period, namely a set of experts, is nonempty.

```
forall p in set dom schedule & schedule(p) <> {}
```

In VDM++, this constraint is added as an *invariant* (abbreviated `inv`) in the instance variables section of the class `Plant`:

```
class Plant
...

instance variables

alarms   : set of Alarm;
schedule: map Period to set of Expert;
inv forall p in set dom schedule & schedule(p) <> {};
end Plant
```

An invariant is a condition that must always hold for an object state. Domain experts have often reported that the ability to record constraints as invariants is one of the most beneficial aspects of modelling. Many of these kinds of constraints go unrecorded or are, at best, implicit. This has significant repercussions for system maintenance and upgrade: Explicit constraints can be checked prior to release; undocumented ones are more likely to be violated.

Guideline 9: Document important properties or constraints as invariants.

2.4.2 Completing Operation Definitions

Having examined the data and recorded important constraints, conditions associated with operations should be considered. Modelling functionality is in some respects more challenging than modelling data. The central question is one of abstraction: "How much detail do I need to give about the algorithm to be used to compute this operation?" Give too much detail and the algorithm may be executable but, like a computer program, it may be too complex to analyse in depth.

The approach taken in VDM++ is to express operations as abstractly as possible in the early stages of design through a technique known as *implicit definition*. An implicit definition of an operation does not give an algorithm, but gives a precondition and postcondition specifying *what* should be computed and under what conditions, not *how* a computation should take place. A precondition records the conditions that must be satisfied for an operation to be correctly applied, i.e., what must hold of the parameters and instance variables before the operation is executed. A postcondition describes the properties that hold after it has been executed, i.e., how the output and instance variables after the operation are related to the inputs and instance variables supplied when the operation was applied. Thus the postcondition is a relation between the initial object state and parameters on one side and the final state and the result on the other.

As an example of implicit operation specification, consider `ExpertToPage`. This takes an alarm and a period as inputs and returns an expert chosen to handle the alarm. An implicit model of `ExpertToPage` includes a precondition stating whether the operation should work for any alarm or period and a postcondition saying what properties must be satisfied by the chosen expert.

`ExpertToPage` has the following signature:

```
ExpertToPage: Alarm * Period ==> Expert
```

However, as indicated earlier, this operation cannot work with just any alarm and period. First, the period must be allocated in the schedule, in order to ensure that there are experts on duty. It also seems obvious to require that the alarm has been registered in the system, i.e., that it is among the possible alarms contained in the instance variable `alarms`:

```
ExpertToPage: Alarm * Period ==> Expert
ExpertToPage(a, p) ==
  is not yet specified
pre a in set alarms and
    p in set dom schedule
```

For example, the operation should not just return any expert who happens to be on duty. The expert should have the correct qualification to cope with the alarm. This is documented in the postcondition:

```
post let expert = RESULT
     in
         expert in set schedule(p) and
         a.GetReqQuali() in set expert.GetQuali();
```

A *let-expression* is used to bind the result of the operation to an identifier `expert` using a special `RESULT` keyword. In the main part of the postcondition, there are two conditions. The first says that the chosen expert is on duty, the second that the qualification requirement is met. Here it has been assumed that access operations `GetReqQuali` and `GetQuali` are defined for the `Alarm` and `Expert` classes respectively. Such access operations must often be defined in object-oriented developments, because of the need to encapsulate the data within objects, limiting the means by which the data may be modified to the designated access operations. Given the pre- and postconditions, the operation has been fully modelled. An executable body for the operation containing details of the selection algorithm can be introduced later.

A full algorithmic description of the function would entail giving an algorithm for selecting the expert, perhaps passing over the database of experts in some particular order, choosing one on the basis of some criterion. Analysing such a model, one would have to work out "retrospectively" what conditions the inputs must satisfy for the operation's invocation to be valid. Note that the requirements do *not* state which of the experts should be chosen if more than one expert with the right qualification is on duty in the given period. The use of an implicit definition allows designers to defer this choice to the latest reasonable point in the development process.

> **Guideline 10:** When there are several alternative ways of performing some functionality, use an implicit definition so that subsequent development work is not biased.

Having provided an abstract description of the expert selection function, one can ask whether it is always possible to find an expert with the right qualification. With the current model this is not yet guaranteed because there is no constraint on the schedule of experts to ensure that there is adequate coverage of all the qualifications that might be required. However, this constraint can be introduced by adding an invariant, as shown here. The expression may appear complex at first sight, but the reader need not examine it in detail at this stage.

```
instance variables

alarms   : set of Alarm;
schedule: map Period to set of Expert;
inv forall p in set dom schedule & schedule(p) <> {};
inv forall a in set alarms &
        forall p in set dom schedule &
            exists expert in set schedule(p) &
                a.GetReqQuali() in set expert.GetQuali();
```

This new larger invariant says that for any alarms and periods registered in the system there must be an expert on duty in the period who has the correct qualification for handling the alarm. This ensures that the specification of `ExpertToPage` makes sense and is implementable, i.e., a desired expert can always be found for all allocated periods. Note that because of the precision of the VDM++ model, detailed questions related to the desired functionality of the system have appeared at an earlier stage of development than might otherwise be the case. For example, the preceding invariant would also be important if the system was extended to allow experts to change their shifts. Note that an invariant like this one also has an impact on the way new periods should be allocated to the system. The invariant identifies a property which is a dominant design parameter for the system but is not identified using a design approach limited to class diagrams and informal text. Again, recording invariants explicitly has downstream benefits for subsequent system maintenance and upgrading.

> **Guideline 11:** When defining operations, try to identify additional invariants.

The operation `NumberOfExperts` has the following signature:

```
NumberOfExperts: Period ==> nat
```

It must return the number of experts on duty in a given period. Probably the period should be known about in the schedule, suggesting a precondition. The result should just be the size (also called the cardinality) of the set of experts on duty in the period. This suggests the operation body:

```
class Plant
...
public NumberOfExperts: Period ==> nat
NumberOfExperts(p) ==
  return card schedule(p)
pre p in set dom schedule;
end Plant
```

In this case, the very simple operation is defined *explicitly*, without a postcondition. There is no need for a postcondition, as it would just be the same as the body, which is already described in a fairly abstract way.

ExpertIsOnDuty has the following signature:

```
ExpertIsOnDuty: Expert ==> set of Period
```

It must always be possible to ask when any expert is on duty so it is necessary to specify a precondition for this operation. Again a postcondition is not necessary because the body of the operation can be described in a high-level and natural way:

```
class Plant
...
public ExpertIsOnDuty: Expert ==> set of Period
ExpertIsOnDuty(ex) ==
  return {p | p in set dom schedule &
              ex in set schedule(p)};
end Plant
```

The body returns the set of all the periods that are in the domain of the schedule and have the given expert on duty. This form of expression may appear unfamiliar to the reader but will be described in much greater detail in later chapters. Such a statement could take up several lines of code in a programming language. In Java, it would look something like:

```
import java.util.*;

class Plant {

  Map schedule;

  Set ExpertIsOnDuty(Integer ex) {
    TreeSet resset = new TreeSet();
    Set keys = schedule.keySet();
    Iterator iterator = keys.iterator();

    while(iterator.hasNext()) {
      Object p = iterator.next();
      if ( ( (Set) schedule.get(p)).contains(ex))
          resset.add(p);
    }

    return resset;
  }
}
```

It is not necessary to understand fully the Java code presented here; the main point is the difference in size between the VDM++ and the Java descriptions. Implementing VDM++ models in Java is treated in much more detail in Chapter 14.

Guideline 12: Try to make explicit operation definitions precise and clear and yet abstract compared to code written in a programming language.

If the (type-checked and consistent) VDM++ model is mapped to the class diagram representation, using the Rose-VDM++ Link, the result is as shown in Figure 2.6.

The class diagram gives a good overview of the model's structure but has no detailed information about the data types used, the constraints (invariants) and the functionality (the operations). This is supplied by the complementary VDM++ view.

2.4.3 Constructing Instances

Before being able to validate the model created so far it is also necessary to consider how to construct instances of the different classes. In the object-oriented world this is typically done using constructors. In VDM++, constructors are simply written as operations with the same name as the class in which they are defined. Typically these operations simply assign actual values to the different instance variables in a class. In this example this definition is made in the Plant class:

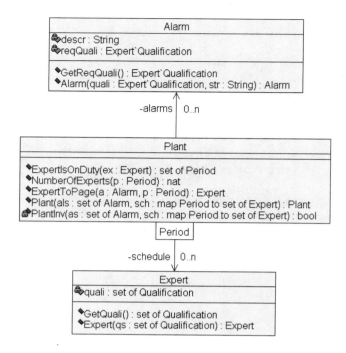

Fig. 2.6: UML class diagram from the VDM++ model

```
class Plant
...
public Plant: set of Alarm *
               map Period to set of Expert ==> Plant
Plant(als,sch) ==
( alarms := als;
  schedule := sch
)
pre PlantInv(als,sch);
end Plant
```

The constructor implicitly returns a reference to the created object, indicated by the class name in the return type of the operation signature. Note that a precondition has been added to the constructor to ensure that it is only applied when the class invariant will hold on the constructed object. This ensures that objects respect the class invariant from the time they are created. The invariant has itself been defined as a separate function, allowing it to be used as both an invariant on the instance variables and a precondition for the constructor. Changes to the invariant need only be recorded in one place (in the function definition), not in both the `inv` part and the constructor:

```
class Plant
...

instance variables

alarms   : set of Alarm;
schedule : map Period to set of Expert;
inv PlantInv(alarms,schedule);

functions

PlantInv: set of Alarm * map Period to set of Expert ->
          bool
PlantInv(as,sch) ==
  (forall p in set dom sch & sch(p) <> {}) and
  (forall a in set as &
      forall p in set dom sch &
         exists expert in set sch(p) &
            a.GetReqQuali() in set expert.GetQuali());
end Plant
```

> **Guideline 13:** Whenever a class has an invariant on its instance variables and it has a constructor, it is worth placing the invariant in a separate function if the constructor needs to assign values to the instance variables involved in the invariant.

For the `Alarm` class the constructor appears as follows:

```
class Alarm
...
operations

public Alarm: Expert`Qualification * String ==> Alarm
Alarm(quali,str) ==
( descr := str;
  reqQuali := quali
);
end Alarm
```

Finally for the `Expert` class the constructor looks like:

```
class Expert
...
operations

public Expert: set of Qualification ==> Expert
Expert(qs) ==
  quali := qs;
end Expert
```

2.5 Validating the Model

The first steps of the construction of a model have been presented. Now one can ask how good the model really is. How can one gain confidence in it? The notion of *validation*, introduced in Chapter 1, involves increasing confidence that a model embodies the properties expected of a system satisfying the client's expectations. In Chapter 13 the different ways of validating VDM++ models will be treated in depth, so this section should mainly be considered an overview. It is worth repeating that this book deals primarily with the use of modelling in design stages when clear specifications are a rarity. The focus is therefore on the validation of models rather than on verification.

Three approaches to validating models are relevant to the technology presented in the book. They are based on integrity properties, systematic testing and rapid prototyping, respectively. *Integrity properties*, also known as *proof obligations*, are formal descriptions of system properties that can be generated automatically using VDMTools. In addition, VDMTools supports validation using conventional testing techniques, including features to enable test coverage documentation directly at the VDM++ level. Finally validation can be made executing models together with other code, e.g., a graphical front end. The interpreter of the VDMTools is able to execute specifications before they are implemented. During execution it automatically checks assertions, i.e., invariants and pre- and postconditions. If some condition does not hold the user is notified with specific information about the violated condition and where the violation occurred. The stronger the conditions are and the more comprehensive the testing is, the higher the confidence in a model will be as a result of the validation.

The rationale for spending time on validation is that the earlier defects can be discovered, the less rework is required and the lower the defect correction cost. Different validation techniques find different kinds of inconsistencies. The cost of applying a validation is to a large degree determined by the quality of the available tool support. Applying the pragmatic approach advocated in this book can be very cost effective, as the case study in Chapter 11 suggests.

2.5.1 Integrity Properties

It is not possible to build a tool that can check the logical consistency of any arbitrary VDM++ model. Indeed, the same is true of checking programs written in expressive programming languages. However, it is possible to generate a number of *integrity properties* for a model. These are expressions that describe the conditions that should hold to ensure internal consistency (discussed in greater detail in Chapter 13). The vast majority of integrity properties are easy to check or refute manually or even automatically. To gain high confidence, each integrity property can be proved mathematically. However, formal proof is beyond the scope of this book (see [Bicarregui&94] for related work in this area).

2.5.2 Automated Validation Using Systematic Testing

VDM++ models may be tested in the same way as programs; this is explained further in Chapter 9. The application of systematic testing on project is also discussed in Chapter 11. It is worth noting that the interpreter supports two modes of working. It is possible to test models interactively, where test expressions are given to the interpreter manually and evaluated immediately. The other mode is a batch mode, where the interpreter is executed automatically on a test suite. In this mode, it is possible to produce test coverage information. This is a very easy way to determine whether all requirements are covered by at least one test case.

To carry out testing it is often necessary to introduce operations for setting the values of the different instance variables. Moreover, the implicitly defined operation ExpertToPage cannot be executed in its current form because it does not have an operation body. However, in this case it is easy to turn the postcondition into a body and make the operation executable:

```
class Plant
...
operations

public ExpertToPage: Alarm * Period ==> Expert
ExpertToPage(a, p) ==
  let expert in set schedule(p) be st
      a.GetReqQuali() in set expert.GetQuali()
  in
    return expert
pre a in set alarms and
    p in set dom schedule
post let expert = RESULT
     in
        expert in set schedule(p) and
        a.GetReqQuali() in set expert.GetQuali();
end Plant
```

This uses a VDM++ *let-be-such-that* expression, which selects an arbitrary expert such that the predicate after "be st" holds. The features from VDMTools supporting this construct are demonstrated in Chapter 3.

2.5.3 Visual Validation Using Rapid Prototyping

The kind of relatively low-level testing introduced earlier is suitable mainly for engineers and less for domain experts, customers, nontechnical staff and management, who might not be familiar with VDM++ or any other object-oriented notations and the details of a model. The VDMTools provide an application programmer's interface (API) to allow interaction with a model via a graphical user interface (GUI) coded in a suitable language, so that customers can test the model directly via the GUI. GUI events are translated into operation calls on the VDM++ model and the return values or state changes are displayed in a form that the user can understand without knowing VDM++ at all. A simple front end for the alarm example could look like that shown in Figure 2.7.

Expert Callout System					
	Expert 1	Expert 2	Expert 3	Expert 4	Expert 5
Period 1	ok		ok	ok	ok
Period 2		ok			ok
Period 3	ok	ok		ok	
Period 4			ok		ok
Period 5		ok		ok	

Expert to page	Number of experts	Expert is on duty

Fig. 2.7: A prototype GUI for the Alarm system

Obviously, better GUIs could be developed for the chemical plant model, but the main point here is that users can easily test the developer's understanding of the requirements of the system. In this way, the VDM++ model is used as an early prototype of the system. The front end can be built using an engineer's favourite tool independent of VDM++. This particular one was developed in Java.

Through an API-based GUI, users can interact with the VDM++ model without prior knowledge of the formalisms used. They can concentrate on setting up scenarios and test sequences in their own terminology. This is a major advantage, which not only reduces the chances of errors found late in the system development but also enhances the client's commitment toward the solution proposed. Experiences have shown that complex (sub) problems found while writing the specification are easily conveyed to the customer using this approach, which in turn will challenge customers to consider their problems in more detail by playing around with the prototype.

The API can also be used to test models together with legacy systems and, in fact, any other code independent of programming language due to its general nature. This is because the API uses the Common Object Request Broker Architecture (CORBA), allowing communication with code written in any language for which an Object Request Broker (ORB) exists. Currently, free and commercial ORBs exist for all major languages. As CORBA is a network-based architecture, this also means that multiple external clients located on a network can communicate with a single instance of VDMTools. Therefore this validation approach can also be used for distributed models; this is illustrated in Chapter 13. A more sophisticated example than the chemical plant GUI is the GUI to the POP3 model, part of which is illustrated in Figure 13.4.

2.6 Generating Code

VDMTools provides automatic code generators that produce directly compilable code. Typically, abstract models like that of the chemical plant need more work before they can have code generated from them and delivered to a customer. Models intended for code generation are typically more design- and implementation-oriented, in particular if there are strong efficiency requirements. However, even models containing high-level expressions like the let-be-such-that introduced in the previous section can have code generated from them in C++ or Java, compiled and run.

In addition to the drastic reduction in the time to construct an implementation, automatic code generation offers a number of benefits. Foremost among these is the strong correspondence between the abstract model and the generated code. This simplifies understanding of the code and its architecture. This also opens up the possibility of reusing test data originally used for testing the VDM++ model. It is possible that the algorithm used in the generated code is inappropriate in a particular instance, but in this case it is possible to generate code skeletons, which may then be hand-implemented. The way of producing the final Java code is presented in Chapter 14.

2.7 Summary

This chapter provided an overview of the process of using the VDM++ notation to express properties about a given problem. The compact but comprehensive example used to illustrate this was an alarm system for which a combination of UML class diagrams and VDM++ was derived. In addition, the VDM++ technology, including the different validation techniques and derivation of code in a programming language, has been presented.

3

VDM++ Tool Support

The aim of this chapter is to introduce the features of the software tools that support the combination of formal modelling in VDM++ and object-oriented design using UML class diagrams. This is done by providing a "hands-on" tour of the tools' functionality using variants of the alarm example introduced in the previous chapter. This chapter should enable the reader to use the software tool support for exercises in the rest of the book. However, it will by no means be a complete tour of the functionality of the tools introduced. Full details are provided in the user manuals accessible via the book's Web site (http://www.vdmbook.com).

3.1 Introduction

One of the main benefits of the approach combining VDM++ and UML class diagrams in this book is the ability to use software tools to assist in the analysis of the models. Often the support for this kind of checking is very limited for UML tools that concentrate on the structural view. However, the combination of Rational Rose® and VDMTools® provides a significant number of features for validation of such models.

This chapter is organised so that it can be read using both Rational Rose and VDM-Tools support or using only VDMTools support if Rational Rose is not available or desired.

Section 3.2 describes how to fetch the tools and associated license files. For those readers who would like to start using Rational Rose, Section 3.3 briefly explains how a first model can be built. Section 3.4 shows all the core functionality of VDMTools. This includes how one can use VDMTools to configure a VDM++ project, perform static checking and synchronise definitions for all classes with a Rose model, if desired. This section also describes validation techniques based on testing and debugging principles using VDMTools. It goes on to show how internal consistency checking facilities can be used to identify potential sources of run-time errors. Finally this section provides a brief introduction to further functionality from VDMTools used later in this book. The tables illustrating the different kinds of buttons from VDMTools all contain a column called "Used" indicating whether the use of the corresponding button is covered by this chapter. The chapter concludes with a short summary.

3.2 Getting Hold of the Software

In order to run the examples and exercises presented in this book it is necessary to get hold of two separate tools. Both of them have setup programs that can be found on the Web, but both of them are license controlled. In order to run the software on your own computer you need to follow these instructions:

VDMTools®: This tool set was originally developed by a Danish company, IFAD A/S. Distribution and licensing arrangements change with time, so you should visit the book's Web site (`http://www.vdmbook.com`) for instructions on how to obtain the most recent installation and getting any associated manuals.

Rational Rose®: This is the commercial version of Rational Rose from Rational, now owned by IBM. It is necessary to either purchase a license for this tool or get a free evaluation license. This tool is under continually being improved so again you should visit the book's Web site (`http://www.vdmbook.com`) for instructions on how to obtain the most recent installation. In the rest of this book we will simply use the term "Rose" to refer to this tool.

When these setup programs are executed they will automatically install the selected tool onto your hard disk. VDMTools has a feature for automatically linking back and forth to Rational Rose called the Rose-Link and that must be installed as described in the relevant manual [UserManPP]. The book's Web pages contain the examples used in this book in both `.rtf` and ASCII `.vpp` formats. These should be downloaded to your hard disk as well, so that you can manipulate them while working through the book. In the rest of this chapter it is assumed that installation has been carried out so that the VDM++/UML tool support is available on your computer.

3.3 Using Rose

This section describes how tool support can be used if one wishes to start the modelling process with UML class diagrams, as shown in that chapter.

The `alarmumlinit.mdl` file can be found on the book's web pages. This UML class diagram model is identical to the initial class diagram from the previous chapter (see Figure 2.2) except that the `Plant` class has been updated with the three operations identified in that chapter. Note that the operations have not yet been given any signatures. Download this `.mdl` file (but **not** the `.rtf` or the `.vpp` files) to your working directory and open it using Rose. When this model is opened in Rose the class diagram should look like Figure 3.1.

In Chapter 2 it was explained that Rose makes associations `public` by default, indicated by the small '+' next to the rôle names. Select one of the associations by pointing to it with the mouse and clicking the left mouse button. While it is selected

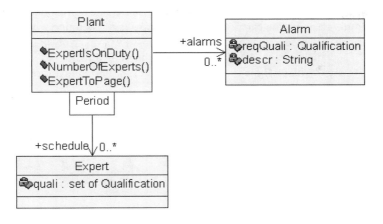

Fig. 3.1: The initial UML class diagram.

click the right mouse button. Select the `private` entry in the drop-down menu. This should now change the plus to a '-' in front of the rôle name. Do the same for the other associations.

Now let us update the signatures for the operations in the `Plant` class. This can be done by double-clicking the class and choosing the **operations** pane. When an operation from the list is double-clicked the information about the selected operation is shown. To describe the signature of the operation the return type must be entered in the **general** pane, and in the **detail** pane one must click the right mouse button in the **arguments** section to insert a new parameter name and its corresponding type.

Add the information about some of the operations from Chapter 2 in this way. Note that, because adding this kind of information at the Rose level requires a lot of clicking and selecting with the mouse, most users prefer to add the signature details at the VDM++ textual level and then have the appropriate information inserted in the correct places automatically.

The changes made to the operation signatures may not be visible directly in your class diagram. This is because Rose has different ways of filtering how much information one would like to have in a diagram. In the **Format** menu it is possible to change how the selected classes are be shown. Here it is the **Show Operation Signature** item that is relevant.

When you would like to move the UML class diagram model to a VDM++ view, you should start the VDM++ Toolbox. Start by saving the project, although it is empty, and give it the same file name (up to the dot) as the Rose model file. Because the `.mdl` file is called `alarmumlinit.mdl` the VDM++ project file should be called `alarmumlinit.prj`. When the Rose Link is invoked from the VDMTools GUI (either by pressing the Rose-Link button (see Table 3.4 later) or by selecting it from the **Windows** menu), a new window will pop up (see Figure 3.2), indicating that at the UML level, three classes have been added (indicated by the **A**, for 'Added'). VDMTools suggests moving them from UML class diagrams to the textual VDM++ representation (indicated with the direction of the small **Action** arrows). Press the **Map** button and a new file for each class gets created in the same directory as the project file and the UML model file. All the files created are present in rich text format (.rtf).

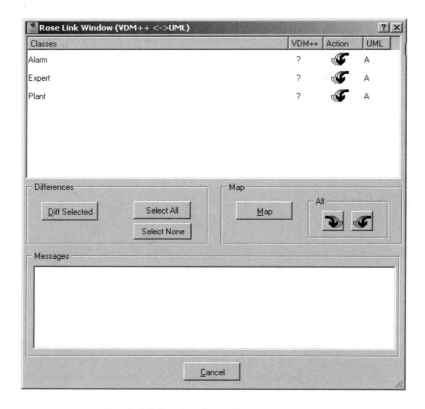

Fig. 3.2: Mapping from UML to VDM++

An overview of the files that have been generated can be seen from the **Manager** window in VDMTools. If you change from the **Project** pane to the **Class** pane in this window, an overview of the different classes that have been syntax checked will be present (see Figure 3.3). The **S**s after each class indicate that the three classes have been successfully syntax checked.

The UML class diagram view of the system will be revisited at the end of the next section, but here the "tour" will continue at the VDM++ level for a while.

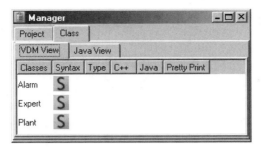

Fig. 3.3: The initial VDM++ class overview

3.4 Using VDMTools

The graphical user interface for VDMTools uses buttons to invoke the major functions. As each area of the tools' functionality is introduced, a table gives the icons on the relevant buttons. It is also indicated whether the buttons are directly referred to in this chapter.

Table 3.1: VDMTools project buttons

Button	Used	Explanation
	No	Create a new project
	No	Load an existing project
	Yes	Save the current project
	No	Save the current project under a new name
	Yes	Add selected files to project
	Yes	Remove selected files from project
	Yes	Show and edit current project options
	Yes	Select tool options

3.4.1 Configuring VDM++ Project

If you followed Section 3.3, a project file called `alarmumlinit.prj` will already have been created. On the book's Web page there is also a file called `Test1.rtf` (it also exists in ASCII format as `Test1.vpp`). Download this file to the same directory as the project file. If you did not follow Section 3.3, download all the `.rtf` files from the `http://www.vdmbook.com/alarminit` directory on the book's Web page to your hard disk. Whether or not you followed Section 3.3, use the **Add selected files to project** button (see Table 3.1) to include all four `.rtf` files in the project. If, alternatively, you do not wish to use `.rtf` format for your VDM++ files, add the `.vpp` files instead. Save the project by pressing the **Saves current project** button (see Table 3.1).

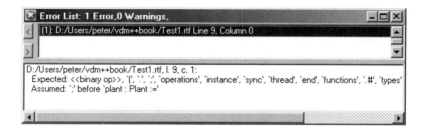

Fig. 3.4: A syntax error

3.4.2 Syntax Checking

Select all the files listed in the **Manager** window and press the **Syntax check** button (see Table 3.2). Three of the files should be syntactically correct, and the `Test1.rtf` file contains one deliberate syntax error. The **Error window** will pop up, as shown in Figure 3.4. In addition the **Source window** will indicate the place in the source file where the error has been located. The error in this case is a typical one; a semicolon separating the different definitions has been forgotten.

3.4.3 Setting Tool Options

This error must be fixed in the correct file (`Test1.rtf`). It is possible to start Microsoft Word manually every time a change is to be made to one of the source files in a VDM++ project. However, it makes sense to configure VDMTools so that it can start it automatically when required. This can be done by pressing the **tool options** button (see Table 3.1). When this is done, a window like that shown in Figure 3.5 will appear. Here the absolute path to your installation of Microsoft Word (or to any other editor that can cope with the source files in the format you have chosen to use) must be inserted.

Table 3.2: VDMTools action buttons

Button	Used	Explanation
SYN-TAX	Yes	Perform syntax check on selected classes
TYPE	Yes	Perform type check on selected classes
INTE-GRITY	Yes	Generate integrity properties for the selected classes
VDM C++	No	Generate C++ code for selected classes
VDM	Yes	Generate Java code for selected classes
pp	No	Pretty print the selected classes
VDM	No	Generate VDM++ classes for the selected Java classes

Table 3.3: VDMTools file buttons

Button	Used	Explanation
	Yes	Open selected files in external editor
	No	Close selected file from source window
	No	Close all files from source window

If this is set, it is then possible to invoke your preferred editor directly on the selected file(s) by pressing the **external editor** button (see Table 3.3). Select the Test1.rtf file and press this button. Add the semicolon that is missing and save the file. Note that the **S**-symbol has changed (a red warning triangle is superimposed) to indicate that the file has been modified and needs to be checked again. When the file is syntax checked again, no errors should be reported.

3.4.4 Type Checking

When all four classes are syntactically correct, select the **Class** pane in the **Manager** window. Here each of the four different classes should now have a small **S** indicating that they have been checked for syntactic correctness. Select all of them and press the **Type check** button (see Table 3.2)

In the **Manager** window all the classes will now get a **T** with a red line through it. This is an indication that type errors have been discovered in all classes. The **Error List**

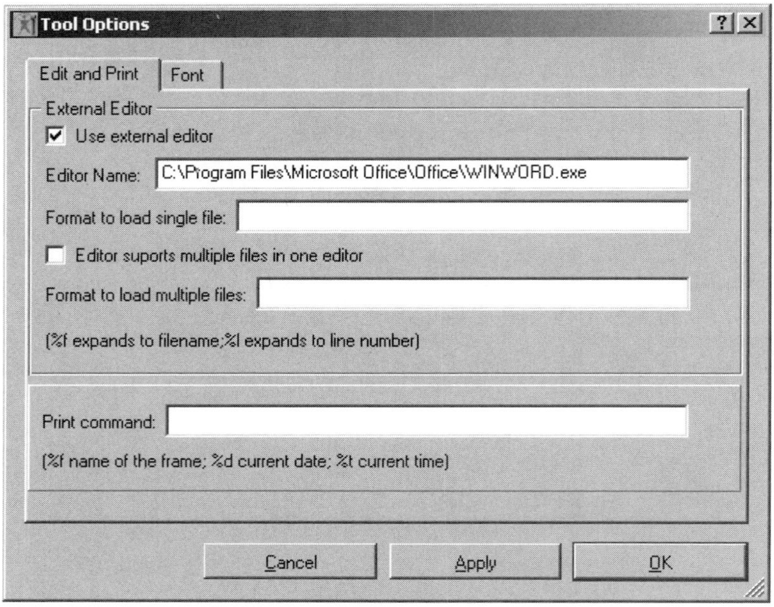

Fig. 3.5: Tool options in VDMTools

window will now appear with a long list of errors; see Figure 3.6. It is possible to move between the different errors by either clicking on a particular error in the upper part of the window or using the $\boxed{<}$ and $\boxed{>}$ buttons on the left-hand side of the window. Whenever the current error in the Error List window is changed, the corresponding position in the Source window is also updated. The first error reported here is that the type String is not defined (it is not a built-in type in VDM). Open the Alarm class in the external editor and add the types keyword before the line with the end Alarm part. After the types keyword type String = seq of char. This is a definition of a type in VDM++ as a sequence of characters. Save the file and look at VDMTools again. Note how the S symbol after the Alarm class has a red warning triangle on top of it. This is an indication that the file has changed on the file system since it was last syntax checked. If you syntax check and then type check again, this error should disappear. In this window, red warning triangles generally indicate that the underlying model files have changed and a check should be performed again; a stroke through the letter indicates that the model failed the check.

Exercise 3.1★ Correct all the errors discovered by the type checker and syntax and type check your corrected files until no syntax and type errors appear. **Hint:** Consult the model presented in Chapter 2 to see how values (note using "=" rather than ":="), types and constructors should be defined and how access modifiers should be used. □

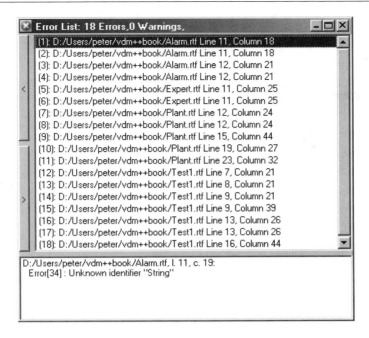

Fig. 3.6: Type errors for initial VDM++ model

3.4.5 Mapping Back to Rose

It is recommended that type errors always be removed before moving a collection of
VDM++ classes up to the UML class diagram level. In a few cases it is impossible
for the Rose-VDM++ Link feature from VDMTools to, for example, distinguish references to classes and references to types when mapping is carried out on a model
with type errors. Once the type errors have been fixed, press the Rose link button (see
Table 3.4) again. Note that it can see that modifications have been made to all three
classes and that the Test1 class has been added. When Map has been pressed, this
new class is placed under Generated classes under the Logical view folder inside
Rose. This class can be dragged onto the Main class diagram and the different classes
can be moved around to improve appearance. Note that the type definitions from the
VDM++ level are not presented at the UML level because there is nothing equivalent
at the UML class diagram level.

Exercise 3.2 Add an operation to one of the classes at the UML level and add
an instance variable at one of the other classes at the VDM++ level. Syntax and type
check the class that have been changed at the VDM++ level and invoke the Rose-
VDM++ Link to update the different views. □

Table 3.4: VDMTools window buttons

Button	Used	Explanation
	No	Allow user to view text
	No	Close selected file from source window
	No	Show project log
	Yes	Call the Rose Link window
	Yes	Open the VDM++ interpreter window
	No	List all errors that occurred
	No	Show integrity properties

3.4.6 Reconfiguring the VDM Project

To avoid having to type in the body of the different invariants, functions and operations presented in Chapter 2, it is possible to reconfigure the files in the project. Remove all the current files by marking them and then press the **Remove File from Project** button (see Table 3.1). Having done this, add all the .rtf files from the alarmfinal directory from the book's Web page (they must be downloaded to the hard disk first). Select all of them and syntax and type check them. No errors should show up this time. If synchronisation with Rose is done, a number of new member definitions will be moved up to the UML class diagram level.

3.4.7 Interpreting VDM++ Models

In addition to the static kind of analysis presented in the previous sections, insight into models can be gained by executing them. This is called *interpreting* the models because they are not compiled but rather are interpreted by an abstract machine. Thus, let us now turn our attention to validating the VDM++ model that has been constructed so far using traditional testing and debugging techniques. Press the button to open the interpreter window (see Table 3.4) and the window shown in Figure 3.7 will appear.

To access the definitions that have been read into the current project, first press the button that **Initialises the interpreter** (see Table 3.5) to initialise the interpreter with all the definitions from the current VDM++ model. Whenever updates have been made to the VDM++ model, syntax checking the newest version does not update the interpreter's internal representation. Thus, whenever new definitions need to be accessed in the interpreter, it is necessary to press this initialisation button again.

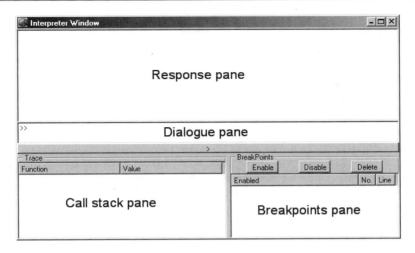

Fig. 3.7: Start up window for the VDM++ interpreter

Table 3.5: VDMTools window buttons

Button	Used	Explanation
	Yes	Initialise the interpreter
	Yes	Perform a step
	No	Step inside
	No	Perform a single step
	No	Continue the execution
	No	Stop the interpreter
	No	Jump to where the current function or operation was called
	No	Jump to where the current subfunction was called
	No	Finish the execution

The top two panes of the tool are, respectively, the **Response** and **Dialogue** panes: you can enter commands directly to the interpreter in the **Dialogue** pane and you receive output from the interpreter in the **Response** pane. In the **Dialogue** pane try to issue a command like:

```
create al:= new Alarm(<Mech>,"Mechanical fault")
```

and press RETURN.

The command `create` is used to make an instance of a class. In this case it makes use of the constructor from the `Alarm` class. The command `print` evaluates an expression. For example, one can see the instance just created by typing:

```
print al
```

In the **Response** pane the instance has a unique object reference and it is possible to see its class (`Alarm`) and the value of its instance variables.

By creating an instance of the `Test1` class it is possible to interpret functionality defined in this VDM++ model. The creation looks like:

```
create test:= new Test1()
```

By inspecting the `test` object using the `print` command it is possible to see that its instance variables themselves are all references to other objects. Because each of them is declared private it is not possible to inspect them directly using the traditional dot notation, e.g., `test.ex1`. Instead we can try to set a *breakpoint* in the operation `Run` defined inside the `Test1` class. This is done as follows:

```
break Test1'Run
```

Now the `Run` operation will appear in the **Breakpoint** pane inside the **Interpreter** window. Now a `debug` command must be used (the only difference from the `print` command is that `debug` breaks the execution if any of the active breakpoints are encountered during the execution):

```
debug test.Run()
```

Interpretation is stopped at the start of the `Run` operation inside the `Test1` class. In the **Source** window this is shown by a small marker indicating the position where the break has been made (see Figure 3.8). In addition, the **Call Stack** pane from the **Interpreter** window is updated with the current call stack. The *call stack* is the stack of active function and operation calls. Now the different interpreter actions can be used. For example, try pressing the **step** button (see Table 3.5) twice to step through the body of the Run operation. Note how it is possible to inspect the current values using the print command, e.g., the periods collection.

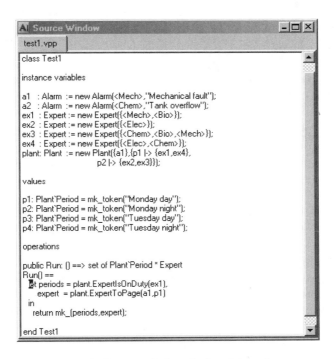

Fig. 3.8: Source window during debugging

Sometimes it is not clear why a function or operation is returning a particular result when applied with some input values. In such cases it is valuable to be able to debug the VDM++ model. The other icons for the interpreter are given in Table 3.5 (all coloured green in the tool) corresponds to different ways of controlling the execution of a given expression in a context where different breakpoints are set. This corresponds to what one can find in programming language debuggers, so this matter will not be elaborated here.

3.4.8 Setting Project Options and Breaking Conditions

Several options can be set in the interpreter. These influence the level of additional checking performed on various features of the model, e.g., preconditions, postconditions and invariants. The project options dialog (see Figure 3.9) can be accessed by the button that **selects tool options** (see Table 3.1).

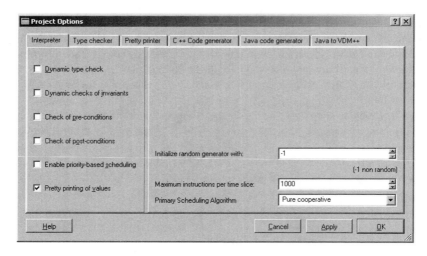

Fig. 3.9: Setting interpreter options

Change the settings for the interpreter options so that precondition checking is enabled. Obtain the previous debug command using the up arrow in the Dialog pane (there is a history of previous commands issued here). When the breakpoint in Run is reached it will be possible to access the instance variables including plant. Instead of continuing the execution of the Run operation try to ask for a new evaluation in the current context. Type:

```
print plant.NumberOfExperts(mk_token("Wednesday"))
```

Because this period has not yet been scheduled, this will break the precondition of the NumberOfExperts operation. In the response pane this message will be issued:

```
Run-Time Error 58: The precondition evaluated to false
```

In addition, the Source window shows in the VDM++ model source file where the problem occured and the function trace shows the call stack, which in this case is only one call deep. By clicking on the three dots in the Call stack pane, the argument of the listed operation call can be expanded. Such dynamic checks of properties can be used to gain more confidence in the correctness of a model.

3.4.9 Integrity Checking

Another way to increase confidence in the internal consistency of a VDM++ model is to use the *integrity examiner* from VDMTools. This extends the static checking capabilities of the VDMTools by scanning through VDM++ models, checking for potential sources of internal inconsistencies or integrity violations. The checks include the violation of data type invariants, preconditions, postconditions, sequence bounds and map domains. Each *integrity property* is presented as a VDM++ expression that should evaluate to true – if it evaluates to false instead, this indicates that there is a potential problem with the corresponding part of the VDM++ model. The development of tools that perform integrity checks statically is a research topic. The only fully general technique for integrity checking is formal proof, but that is beyond the scope of this book.

By pressing the **integrity checking** button (see Table 3.2) when all the classes are selected, an **Integrity properties** window will appear (see Figure 3.10).

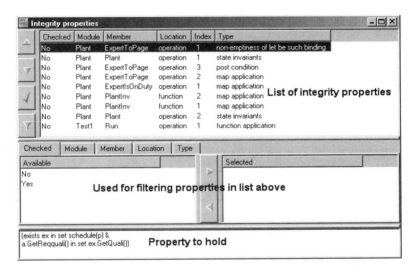

Fig. 3.10: Integrity properties for the `Alarm` example

The **Integrity properties** window shown in Figure 3.10 is divided into three areas. The area at the top lists the integrity properties currently in view. Initially this will show all the integrity properties for the classes that have been checked. To the left of the list are four buttons. The first two are for scrolling through the list. The third marks the currently highlighted property as having been checked manually. The fourth button activates a filter that can be used to limit the view to certain kinds of integrity property. The second part of the window is used to define the filtering criteria. The bottom part of the window shows the currently selected property. This property should hold at the given position in the **source** window.

Integrity checking can be valuable in the late stages of a model's development. Checking each property can help to identify potential run-time errors. The use of this feature will be taken up again in Chapter 13.

3.4.10 A Command-Line Interface

So far, only the graphical user interface of VDMTools has been presented. In addition, there is a command-line interface that is useful when one wishes to test a VDM++ model systematically and derive data on the extent to which the tests have exercised all parts of the model (see Chapter 9 for more information).

The command-line version of VDMTools is executed by typing vppde (shorthand for the VDM++ development environment) in a DOS or UNIX shell where the path to this binary is included or the PATH environment variable is set to include the directory in which the executable is placed. This command takes various options, for example, the "-t" flag activates the type checker and the "-i" flag activates the interpreter. Hence, the preceding specification is type checked by typing

```
vppde -t Plant.rtf Alarm.rtf Expert.rtf
```

This command invokes the type checker for the three files listed as parameters.

The command-line interpreter requires that a *test argument file* is developed. This can contain a list of expressions to be evaluated by the interpreter. The interpreter can take more flags for further checking, and it also needs an argument file (normally with a .arg extension). An example could be the following file, which is called test1_1.arg and runs a version of the preceeding test script:

```
new Test1().Run()
```

This uses a new class Test1:

```
class Test1

instance variables

a1    : Alarm  := new Alarm(<Mech>,"Mechanical fault");
a2    : Alarm  := new Alarm(<Chem>,"Tank overflow");
ex1   : Expert := new Expert({<Mech>,<Bio>});
ex2   : Expert := new Expert({<Elec>});
ex3   : Expert := new Expert({<Chem>,<Bio>,<Mech>});
ex4   : Expert := new Expert({<Elec>,<Chem>});
plant: Plant   := new Plant({a1},{p1 |-> {ex1,ex4},
                                   p2 |-> {ex2,ex3}});

values

p1: Plant`Period = mk_token("Monday day");
p2: Plant`Period = mk_token("Monday night");
```

```
p3: Plant'Period = mk_token("Tuesday day");
p4: Plant'Period = mk_token("Tuesday night");

operations

public Run: () ==> set of Plant'Period * Expert
Run() ==
  let periods = plant.ExpertIsOnDuty(ex1),
      expert  = plant.ExpertToPage(a1,p1)
  in
    return mk_(periods,expert);

end Test1
```

Then the following command is executed:

```
vppde -iDIPQ test1_1.arg Plant.rtf Alarm.rtf Expert.rtf
          Test1.rtf
```

While testing from the command line, the interpreter can collect test coverage information (using an -R flag). This requires that we first create an empty test coverage file (called vdm.tc) as follows:

```
vppde -p -R vdm.tc Plant.rtf Alarm.rtf Expert.rtf
       Test1.rtf
```

and then execute a test case based on the preceding argument file:

```
vppde -iDIPQ -R vdm.tc testing/test1_1.arg Plant.rtf
                        Alarm.rtf Expert.rtf Test1.rtf
```

The produced output file contains the expected result:

```
mk_({ mk_token(
  "Monday day" ) },
  objref4(Expert):
  < Expert'quali = { <Bio>,<Mech>} } > )
```

The resulting test coverage information is displayed in the pretty-printed specification on the book's Web page.

The vppde commands can be called from batch or shell script files under Windows and Unix. This approach is recommended for systematic testing and is illustrated in Chapter 9.

3.4.11 CORBA-Based API

VDMTools provides a CORBA (Common Object Request Broker Architecture) compliant application programming interface (API), which allows other programs to access a running Toolbox. This was also explained in Section 2.5.3. This enables external control of Toolbox components such as the type checker, interpreter and debugger. The API allows any code, such as a graphical front end or existing legacy code, to control the Toolbox. Chapter 13 shows how this feature can be used to validate that the model is behaving as the user, who may not be familiar with VDM++ or UML, expects.

3.4.12 Code Generation Features

VDMTools supports automatic generation of C++ and Java code from VDM++ models, helping to achieve consistency between specification and implementation. This was also explained in Section 2.6. The code generator produces fully executable code for 95% of all VDM++ constructs, and there are facilities for including user-defined code for nonexecutable parts of the specification. Once a specification has been tested, the code generator can be applied to obtain a rapid implementation automatically. In Chapter 14 it will be illustrated how this feature may be used to produce a Java implementation of a VDM++ model.

3.5 Summary

This chapter has introduced the following major features of tool support for VDM++:

- using Rational Rose with class diagrams;
- mapping between Rose and VDMTools;
- configuration of selected VDM++ files;
- syntax checking of VDM++ models;
- type checking of VDM++ models;
- the notion of error messages;
- executing and debugging VDM++ models;
- a command-line interface;
- pretty printing VDM++ models with test coverage information;
- the existence of integrity checking;
- the existence of the application programming interface;
- the existence of code generators; and
- setting tool and project options.

Part II

Modelling Object-oriented Systems in VDM++

4

Defining Data

This chapter and the next aim to provide a practical introduction to system modelling in the VDM++ language and to equip the reader for the more sophisticated modelling techniques described in later chapters. Here, the focus is on the elements of data modelling; the next chapter provides a working introduction to the description of functionality. The language components introduced in this chapter are described in more detail in the full language manual on the book's Web site.

4.1 Introduction

Although the approach to system modelling advocated in this book is quite generally applicable, it is worth considering its implementation in a particular language in order to provide a concrete basis for illustrating its more advanced features and to give the reader an opportunity to experiment. It is therefore necessary to devote a couple of chapters to basic language detail before moving on to the more sophisticated aspects of modelling. This chapter and the next therefore introduce, in some detail, the basic elements of VDM++. This includes the language features, but it also covers the basic rules and checks on well-formedness, consistency and type correctness that underpin a formal language such as VDM++.

Object-oriented models place data and functionality together within objects that conform to class descriptions. This chapter gives an overview of the format in which VDM++ classes are described (Section 4.2) and defines the fundamental data types used for describing constants and variables (Sections 4.3 to 4.5). These basic types are quite abstract, compared to those in programming languages; the capacity for modelling concepts closer to the level of code comes to the fore more in Chapter 5, where the range of styles of function and operation definition are presented.

A Running Example: Electronic Patient Records (EPR)

A simple running example will be used in this chapter and the next to illustrate most of the features of VDM++ in depth. Those features not naturally covered in the example will be illustrated in later chapters. The example concerns the generation and storage of Electronic Patient Records (EPR) for a health service provider and is inspired by the Danish national registration system.

In the example system, as in Denmark, everyone has a unique identifier called their CPR (Central Population Register) number, used for the delivery of a range of services, including health care. The CPR number is a ten-digit code: the first six digits represent the person's date of birth (day-month-year); the remaining four digits are uniquely generated and encode the person's sex. For the delivery of health care, the model deals with *patients*. Each patient has a name, a CPR number and a history in the form of a series of events in which the person had some interaction with the health care provider. When a baby is born, a unique new CPR number has to be generated.

It is not the purpose of this chapter to illustrate a design methodology; instead, the focus is on the basics of the modelling language. However, this brief description suggests several elements of the model. There will be classes representing dates, CPR numbers, the CPR registry itself, and patients. In this chapter, the example will mainly be used to illustrate basic data ideas: the instance variables of the objects mentioned, their types and invariants. Functionality, such as the generation of CPR numbers and the recording of events, is dealt with in the next chapter.

Lexical Issues

VDM++ is a case-sensitive language. For example, the keyword Class is not a valid keyword, whereas the keyword class is, and the identifiers HELLO and hello are not the same. On a practical note, in VDMTools the character set allows Unicode [Unicode] characters, thus enabling the use of identifiers with non-ASCII characters.

4.2 Object-oriented Structuring

A VDM++ model consists of a collection of *class descriptions*. A class represents a collection of objects that share common elements such as attributes or operations. A class description provides a template for objects created in the class, defining all the common elements. Some of these elements will be visible to the world outside the created object (the other objects in the system), and some of them will remain internal to the object itself.

The structure of a class description is presented in Figure 4.1. A class is introduced by the keyword class, followed by a class name. The description consists of several blocks, each preceded by a keyword indicating the kind of element described in that block. Any element defined in a block is referred to as a *class member*. The following kinds of block are possible:

- *Instance variables*, which model the internal state of the object (see Section 4.3).
- *Types*: definitions of data types (see Section 4.4).
- *Values*: definitions of constants (see Section 4.5).
- *Functions*: definitions of pure functions that do not affect the instance variables (see Chapter 5).

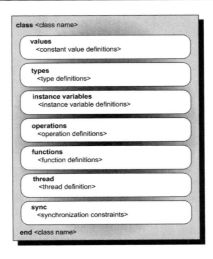

Fig. 4.1: Structure of a VDM++ class definition

- *Operations*: definitions of operations that can modify the instance variables (see Chapter 5).
- *Threads*. Objects in VDM++ can be either active or passive. When an object is active, it has a thread that defines the activities an object will carry out when the thread is started. Threads are introduced in Chapter 12.
- *Synchronisation constraints*. VDM++ supports modelling of concurrent software systems, meaning that objects may exist concurrently and act in parallel. This block specifies the constraints on communication between concurrent objects is constrained. Synchronisation constraints are described in detail in Chapter 12.

In a modelling language such as VDM++, geared to analysis and validation, ease of reading and comprehension are of paramount importance. Consequently, there are relatively few restrictions on the order in which parts of a model may be presented. A class description can contain any number of blocks of any kind and in any order. Each block may contain any number of definitions, and the order of the definitions is not generally constrained. Within each block definitions are separated by semicolons. The final definition in a block may be followed by an optional semicolon.

In the running example, consider a simple class to model dates. A date will have instance variables representing day, month and year, and data types to represent the values that can be taken by the instance variables. Thus, the class definition would have an instance variables block and a types block. The outline structure of the class definition in VDM++ will ultimately be as follows:

```
class Date

instance variables

  day   : Day
  month: Month
  year  : Year

types

  Day   = ...
  Month = ...
  Year  = ...

end Date
```

Where inheritance relationships exist between classes, the inheritor is called the *subclass* and the class from which it inherits is called the *superclass*. The subclass definition is, in effect, extended with the members of the superclass that the superclass has indicated are available for inheritance by having them labelled *public* or *protected* (see Section 4.2.1). A class can inherit from more than one superclass. The only restriction on such *multiple inheritance* is that at most one of the superclasses may have a thread definition.

An inheritance relation is indicated by the keyword is subclass of after the name of the subclass, followed by a list of the names of its superclasses, separated by commas:

```
class class name is subclass of sc1, sc2, ..., scn

...

end class name
```

Exercise 4.1 Define a class hierarchy in VDM++ in which the concepts *animals*, *birds* (e.g., ducks, penguins and eagles) and *fish* (e.g., cod and salmon) are related to one another using simple inheritance (no multiple inheritance). □

Exercise 4.2 Consider again the concepts introduced in Exercise 4.1. Thinking about the functionality of the different classes, do you think the inheritance structure is appropriate? If, for example, flight is to be modelled as functionality, where should it go in the hierarchy? How could a more appropriate hierarchy be created using multiple inheritance? □

If several superclasses define a class member with the same name, a name conflict occurs. In VDM++ this can be resolved by using *name qualification* in which the name of the member is prefixed by the name of the intended class:

```
class name`definition identifier
```

This notation can also be used for public members (see next subsection). If, for example, one wishes to refer to a type called Sex in a class called Patient it looks like:

```
Patient`Sex
```

Once a class has been defined, it is possible to model the creation of objects belonging to the class using a *new expression*. It is written like this:

```
new class name ()
```

The expression returns a *reference* (see Section 4.4.2) to a newly created object of class *class name*. For example, if an instance of the Date class is required, it will simply be written as new Date(). An object can refer to itself using the keyword self.

VDM++ incorporates the notion of *dynamic binding*, meaning that an object declared to be of a class can be assigned a value (or instantiated) from an object from one of the subclasses of that class.

4.2.1 Visibility Rules

Visibility and scope rules define the points in a model at which particular constructs may be used. The basic rules for VDM++ are similar to those found in many object-oriented programming languages. Visibility can, however, be influenced by using *access modifiers*. Each member declaration inside a class can have an access modifier indicating the visibility of that definition. Three different kinds of access modifiers are available in VDM++: private, protected and public. The level of visibility is determined by *access modifiers* as follows:

Public: A public member is visible to all objects of all classes in the model. Public attributes and operations are indicated in the class diagram by a tilted rectangle, whereas public associations are indicated by a "+" to the left of the association name.

Protected: A protected member is visible only to the class in which it is defined or to its subclasses. Protected attributes and operations are indicated in the diagram by a key symbol, whereas protected associations are indicated by a "#" to the left of the association name.

Private: A private member is visible only in the class in which it is defined. Private attributes and operations are indicated in the diagram by a padlock symbol, whereas private associations are indicated by a "−" to the left of the association name.

If no access modifier is explicitly stated for a construct, the default is `private`.

It is sometimes desirable to be able to use a member, such as a constant value or a defined data type, without reference to a specific instance of the class in which the member is defined. Such members are referred to as `static`. Statically declared constructs are accessed using the class name instead of a reference to an instance. Thus, if an operation called `Op` is defined to be static in a class called `C` it will be referred to as `C'Op` using the mechanism for name qualification. A static class member is not allowed to refer to any nonstatic class member. By default, value and type definitions are all static.

4.3 Instance Variables and Invariants

The internal state of an object is modelled by means of instance variables. The instance variables definition block in a VDM++ class description has the following format. Any number of instance variable definitions may be given, separated by semicolons:

```
class class name

instance variables
instance variable definition 1;
instance variable definition 2;
...
instance variable definition n

end class name
```

Each instance variable definition has the following general form if the variable is not initialised:

```
identifier: type
```

If initialisation is included, this is done in the definition:

```
identifier: type := value
```

The *value* can be any expression (see Section 5.4) that evaluates to a value of type *type*. It yields the initial value of the instance variable that will be supplied when an object of this class is created.

Sometimes the initial value for an instance variable depends on the circumstances at the time the object is created. In this case, the initial values can be supplied at object creation time if a *constructor* (see Section 5.3.4) has been defined for the class. The object creation is written as follows:

```
new class name (value 1, value 2, ..., value n)
```

If initial values are assigned by both a declaration and a constructor, the constructor takes precedence. If no initial value is given at all, the initial value is simply undefined.

Invariants on Instance Variables

As illustrated in the chemical plant example in Chapter 2, one of the most useful features of a formal modelling language such as VDM++ is the support for explicitly recording constraints on the values that can be held in instance variables. There are many situations in which it is necessary to limit the values in instance variables or to say that some important relations must always hold between them. Such constraints are called *invariants*. An invariant definition directly follows the definition of the instance variables themselves as shown here:

```
class class name

instance variables
instance variable definition 1;
instance variable definition 2;
...
instance variable definition n;
inv expression using the instance variables

end class name
```

For example, consider the class definition modelling dates in the EPR system. Each date is constrained so that the day number falls within the range permitted by the month. This could be recorded by an invariant restriction as follows:

```
class Date

instance variables

  day   : nat1;
  month: nat1;
  year  : int;
  inv day <=31 and
      month <= 12 and
      if month in set {4,9,6,11}
      then day <= 30
      else (month = 2) => (day <= 29)

end Date
```

Here, nat1 stands for the type of nonzero natural numbers, and int is the type of integers, both of which are introduced properly in Section 4.4.1. The invariant property is required to hold at all times in any instance of this class. Any attempt to assign a new value to the day or month should be accompanied by a check that the assignment will respect the invariant. By making constraints explicit in this way, VDM++ opens up the possibility of checking systematically that they will always be respected. Invariant checking is often an afterthought in conventional design and programming approaches, whereas it is integral to formally based techniques.

The meaning of the in set operator should be clear from the context here: It has been seen in Chapter 2 and is introduced properly when sets are considered in depth in Chapter 6. Notice that the invariant contains some restrictions that are specific to single-instance variables (e.g., the day must not exceed 31), and some that relate two or more instance variables (e.g., if the month is 2, the day must not exceed 29). A complex invariant describing the leap year requirements would fall into this latter category. This question of placing invariants is revisited with the EPR example after basic types and type definitions have been introduced (Section 4.4.5).

Exercise 4.3 Sketch a class definition modelling nonsquare rectangles. Each rectangle is represented by its height and width as instance variables. □

Exercise 4.4 Suppose you are asked to model a security management system for a shopping mall. The mall consists of *units*, some of which are shops. Each unit has a door with a sensor telling us whether it is open or closed. Each shop has a key lock operated by the owner telling us whether the shop is open for business or not. In addition, each shop has a security alarm that is either in a silent or ringing state. If the shop door is open but the shop is not open for business, the alarm should be sounding. Give outline definitions of classes representing units and shops. Define instance variables for your classes representing units and shops sufficient to model these properties (use the boolean type bool for the states of the shops, doors and alarms). Formulate an

Table 4.1: Overview of the VDM++ basic types

Type	Values
bool	true, false
nat1	1, 2, 3, ...
nat	0, 1, 2, ...
int	..., -2, -1, 0, 1, ...
rat	..., -1/7, -1/356, ..., 1/3, ...
real	..., -12.78356, ..., 0, ..., 1726.34, ...
char	'a', 'b', ..., '1', '2', ... '+', '-' ...
quote	<RED>, <CAR>, <QuoteLit>, ...
token	mk_token (...)

invariant in natural language. □

4.4 Types and Type Definitions

As with most specification and programming languages, the concept of a data type, usually referred to simply as a *type*, is central to VDM++. A type consists of a collection of values along with operators that may be applied to those values. For example, the *integer* type is provided by virtually all programming languages; it consists of a range of whole number values and operators such as addition, subtraction, multiplication, etc.

In VDM++ types fall into two categories:

- *Basic types*, which simply model collections of values, and
- *Constructed types*, which are types resulting from applying a *type constructor* to other types (constructed or basic).

This section introduces the basic and constructed types available in VDM++, using simple examples and the running EPR example. As in a programming language, these basic and constructed types can be used in the definition of new types that are directly relevant to the model. The section ends by showing the format of type definitions and how *invariants* can be used to express constraints on the values of a data type.

4.4.1 Basic Types

The built-in basic types in VDM++ model atomic values (values that cannot be broken down further into component parts). Table 4.1 gives an overview. Most of the basic types are quite well known from programming languages and need little introduction. Some of them are, however, rather more abstract than their programming counterparts and deserve a slightly more detailed introduction.

The basic boolean type bool contains the usual values true and false. Booleans have the usual operators familiar from programming (not, or, and), a logical implication "if ... then ... " operator (=>) and biimplication "if and only if" operator (<=>). Full definitions can be found in the Language Manual.

The numeric types in VDM++ are unbounded: There is no specified maximum natural number or minimum integer, for example. This is simply a reflection of the abstract character of the language as a modelling, rather than a programming, medium. As one might expect, the unboundedness has to be compromised in the tool support. The following relation holds on the numeric types: nat1 ⊂ nat ⊂ int ⊂ rat ⊂ real where "⊂" means subset inclusion. VDM++ has no explicit means of casting one type to another.

Although types can be unbounded, the values within types are finite. For example, the string type (written seq of char) contains all possible finite strings of characters. This finiteness allows operators to be readily applied to values, for example, the len operator will return the length of a string. Because strings are used so frequently there is also a special shorthand notation used for string literals. A string literal is shown with double quotes, e.g., "string". This is a shorthand for a sequence of characters that would have to be written in a more cumbersome way as ['s','t','r','i','n','g'] using the conventional sequence notation. Sequences are introduced in greater depth in Chapter 7.

Two of the basic types deserve special attention: the *quote type* and the *token type*.

A quote literal is a single named value, its name being enclosed between < and >. A quote type has the same name as the quote literal and contains just the quote literal as its only value. So, for example, a quote type that represents the operational "on" value of some device can be modelled as:

```
<On>
```

Quote literals and types are often used in defining enumerations (see Section 4.4.3)

The token type is, in a sense, the most abstract type in VDM++. It represents values that have no internal structure at all. It is used to represent values that require identification but for which the exact representation is irrelevant to the model. For example, suppose a model is being built of an online booking system for a hotel chain. If the purpose of the model is to assess throughput of transactions, one may care that transactions contain customer identifiers but may not be concerned about whether those identifiers are numeric or character data or even how long they are. In such a model, token would be an appropriate type for the customer identifiers.

The only operators available for quote and token types are equality (=) and inequality (<>). The purity of the token type's abstraction has to be compromised in tool support, however. If a model is to be animated or tested, it is often necessary to construct token values and give them names. This is done using a mk_token (read as "make token") constructor. Thus the expression mk_token("fred") means a structureless value called "fred".

The operators available on the base types are summarised in Table 4.2. The right-hand column gives the *type signature* of the operator. To the left of the arrow are listed the types of the operands in order; to the right is shown the type of the result. Some

Table 4.2: Operators on basic types; "T" stands for any type from Table 4.1

Operator	Name	Type
not b	Negation	bool → bool
a and b	Conjunction	bool * bool → bool
a or b	Disjunction	bool * bool → bool
a => b	Implication	bool * bool → bool
a <=> b	Biimplication	bool * bool → bool
a = b	Equality	T * T → bool
a <> b	Inequality	T * T → bool
-x	Unary minus	real → real
abs x	Absolute value	real → real
floor x	Floor	real → int
x + y	Addition	real * real → real
x - y	Difference	real * real → real
x * y	Product	real * real → real
x / y	Division	real * real $\xrightarrow{\sim}$ real
x div y	Integer division	int * int $\xrightarrow{\sim}$ int
x rem y	Remainder	int * int $\xrightarrow{\sim}$ int
x mod y	Modulus	int * int $\xrightarrow{\sim}$ int
x**y	Power	real * real → real
x < y	Less than	real * real → bool
x > y	Greater than	real * real → bool
x <= y	Less or equal	real * real → bool
x >= y	Greater or equal	real * real → bool

operators work on all the basic types; in their table entries the symbol "T" stands for any basic type.

Most operators produce defined results for all possible inputs and are termed *total operators*. It is important, however, to note that some operators are *partial*, meaning their results are undefined for some inputs. For example, division by zero is undefined, so the division operator / is partial. In Table 4.2, and in all tables listing operators in this book, a partial operator is indicated by the use of a modified arrow $\xrightarrow{\sim}$. It is important to protect applications of partial operators (e.g., by invariants or preconditions) to ensure that they are only used within their domains of definition.

4.4.2 Constructed Types

The basic types are rather limiting. It would be difficult to model any system without an ability to represent structured data, such as records, and collections of data, such as sequences. Types representing structured values and collections are built from the basic types by means of *type constructors*. The "workhorse" constructors used to build structured values (unions, products, records, optional types and reference types) are introduced in this section. The type constructors used to model collections, specifically sets, sequences and mappings, are important enough to warrant chapters of their own: Chapters 6, 7 and 8.

Union Types

A union type is formed from several component types and contains all the values from each of the components. Normally, the types participating in a union are *disjoint* (they have no members in common). It is possible to form a union from types that are not disjoint (e.g., `rat` and `real`), but this is considered bad practice.

A union type is represented as follows:

```
type 1 | type 2 | ... | type n
```

The type operators available are listed in Table 4.3, where A stands for the union type. As an example, consider the union type

```
nat | bool
```

The values 5 and `false` are both valid values of this union type.

Table 4.3: Union type operators; "A" stands for any union type

Operator	Name	Type
t1 = t2	Equality	A * A → bool
t1 <> t2	Inequality	A * A → bool

Probably the most common use of type union is in forming *enumerated types* which consist of several distinct (enumerated) values. In VDM++, enumerated types are formed by applying unions of quote types. For example, the type

```
<Green> | <Amber> | <Red>
```

could model the status of a traffic light. The same technique was used in the chemical plant example of Chapter 2 to model qualifications. In the EPR, the sex of a patient could be modelled by the type union

```
<Male> | <Female>
```

A further common use of union types is in combining values of different quote types with record types, for example, in defining recursive data structures with quote literals at the leaves and record structures at each internal node. Exercise 4.6 provides an example of this.

Exercise 4.5 In Exercise 4.4 the state of the door of a unit in the shopping mall was modelled using a Boolean type. What other VDM++ type can be used to model this status? □

Product Types

While a union type combines values drawn from several other types, a product type brings values together into composite structures called *tuples*. In VDM++, a product type consists of a fixed combination of other types. It is written as

```
type 1 * type 2 * ... * type n
```

A tuple is written using the keyword mk_ (read as "make", also called the *tuple constructor*), followed by an enumeration of the various values within parentheses:

```
mk_(value 1, value 2, ..., value n)
```

A product type is most useful in cases where the fields are not homogeneous. For example, a model of addresses might use type definitions of the following form:

```
nat1 * (seq of char) * PostalTown
```

where, for example, PostalTown could be defined as:

```
PostalTown = <London> | <Manchester> | <Newcastle>
```

The operators defined on product types are component selection, equality and inequality (Table 4.4).

Table 4.4: Operators on product type $T=T1*\ldots*Ti*\ldots*Tn$

Operator	Name	Type
t.#i	Select	$T * nat1 \xrightarrow{\sim} Ti$
t1 = t2	Equality	$T * T \rightarrow bool$
t1 <> t2	Inequality	$T * T \rightarrow bool$

Tuple components may be accessed using the select operator. The fields are numbered left to right, starting with 1. Thus, the month can be extracted from a date by selecting field 2:

```
mk_ (12,"Abstraction Avenue",<Manchester>).#2
```

which will yield the string `"Abstraction Avenue"`. The select operator is partial because it is only defined for the the numbers 1 up to the number of fields in the tuple value. Equality and inequality are defined on tuples. Two tuples are equal if and only if they are equal in all their component fields. Otherwise they are unequal.

Record Types

A record type, like a product type, consists of tuples of elements from a fixed combination of other types. With record types, however, the individual components of tuples can, optionally, be named. The general form of a record type definition is as follows

```
tag ::
    field 1 : type 1
    field 2 : type 2
    ...
    field n : type n
```

where the *field* identifiers are optional. Values in a record type are *tagged* with a label giving the type name. Just as with product types, a record constructor is available for constructing values of a record type:

```
mk_tag (value 1, value 2, ..., value n)
```

For example, addresses could be represented by the type:

```
Address ::
    house  : HouseNumber
    street : Street
    town   : PostalTown
```

As suggested in Section 2.3.2, such a record type definition is preferred to a full class definition when only simple access functionality is required, rather than a suite of methods.

The operators on records are presented in Table 4.5, where A is a record type containing a field named i of type Ai. For example, the selection of the street field from an Address record is expressed as follows:

```
mk_Address(15,"The Grove",<London>).street
```

which yields "The Grove".

Table 4.5: Record type operators

Operator	Name	Type
r.i	Field select	A * Id → Ai
r1 = r2	Equality	A * A → bool
r1 <> r2	Inequality	A * A → bool

Exercise 4.6 Give a definition for a data type that represents a binary tree-like structure. A Tree is either a natural number or a Node. A Node contains a natural number and two subtrees, called the "left" and "right" subtrees. □

Optional Types

The optional type constructor takes any type as its argument and adds a special value to it: the value nil.

An optional type is written as

```
[type]
```

For example, the type

```
[bool]
```

has three values: true, false and nil.

In the EPR example, it is important to note that a patient might be nameless. In particular, a baby could have a CPR number as an independent person, but not yet have a name. The data definitions in the Person type can be extended to deal with this by making the patient name type optional.

```
[seq of char]
```

Note that there is an important difference between the empty string, which is of type seq of char, and the special nil value. In many modelling contexts, it is important to differentiate between empty data, e.g., a name in the process of being entered, and absent data or an error result, often modelled by nil and the use of the optional type constructor.

Exercise 4.7 Extend your answer to Exercise 4.6 to allow your tree structure to be empty and to allow nodes to have either or both of their subtrees empty. □

Object Reference Types

An object reference type contains values that identify objects from a given class, serving as references to them. An object reference type is indicated by a class name:

```
class name
```

The *class name* in the object reference type must be the name of a class defined in the model.

Table 4.6 lists the operators available on object reference types. Note that equality in this case refers only to the object references, and not, for example, to the internal states of the objects pointed to by the references.

Table 4.6: Overview of object reference type operators

Operator	Name	Type
t1 = t2	Equality	A * A → bool
t1 <> t2	Inequality	A * A → bool

The most common use of object references is for recording *clientship relations* between classes. This arises when an instance variable (see Section 4.3) in a class, say class B, is declared to be of type A, where A is the name of another class in the model. A is thus an object reference type. B is a client of A in the sense that B may now use the services of A, creating and using objects and invoking their operations.

```
class B

instance variables
   a: A := new A ()

   . . .

end B
```

When an object of class B is created, an object of class A is automatically created. The B object will contain an instance variable a, which is an object reference to the object of class A, and it will be possible to refer to this object using that instance variable.

4.4.3 Type Definitions

In VDM++, a *type definition* associates a *type name* with a type. The use of type definitions with meaningful names is good practice, but it is especially useful in situations where the defined type is needed in several places. Writing a type definition then assures that, whenever the type needs to be changed, this change only needs to be made at one place in the model.

A type definition block starts with the keyword `types`, followed by an arbitrary number of type definitions separated by semicolons (`;`):

```
class class name

types
type definition 1;
type definition 2;
...
type definition n

end class name
```

where a type definition has the form:

```
type name = type
```

or

```
record type
```

For example, the definition of a type that models the discrete values that a traffic light may signal can be written as:

```
class class name

types
TrafficLight = <Green>|<Yellow>|<Red>

end class name
```

The *tag* of a record type is also its type *name*. So

```
class class name

types
Coordinates :: x: nat
               y: nat

end class name
```

is an example of a type definition associating the type name Coordinates with a record type. This has the effect that all record values of this type will be tagged with that name, i.e., mk_Coordinates(10, 20) will be such a value.

In VDM++, comparison of types is based on the collection of values that a type represents, not on the name of that type. For example, consider two types T1 and T2, where T1 is defined as

```
T1 = real
```

and T2 is defined as

```
T2 = real
```

Then values of T1 and T2 may be used interchangeably in expressions, so, for example, the multiplication expression

```
e1 * e2
```

where e1 is of type T1 and e2 is of type T2, is a valid expression in VDM++, whereas it would not be in a strongly typed language, such as Ada.

Exercise 4.8 Think of a way to emulate strong typing in VDM++, for example, defining two different types of natural numbers, the values of which can nevertheless not be compared. □

4.4.4 Invariants on Type Definitions

In much the way that invariants can be defined over instance variables (see Section 4.3), the same can be done for type definitions. An invariant on a type is a boolean expression (called a *predicate*) restricting the type to contain just those values satisfying the predicate. A type invariant takes the following form:

```
type name = type
inv pattern == predicate
```

The *pattern* can be a variable standing for a value drawn from the *data type*. For example, the type of natural numbers less than 100 is expressed:

```
SmallNat = nat
inv s == s < 100
```

In the case of record types, the pattern can be a record pattern with variables for the fields in the record (see Section 5.5 for more details about patterns). For example, consider a record type T, with two fields of natural values, the first of which should always be larger than the second; this can be written as:

```
T :: field1: nat
     field2: nat
inv mk_T(f1, f2) == f1 > f2
```

4.4.5 Types in the EPR Example

Types have been used extensively in this and preceding chapters. However, it is worth showing the format of some type definitions using basic types in the EPR example. Recall the Date class. A simple form of the class definition would provide separate type definitions for days, months and years, allowing these types to be used in any classes that use the Date class. The class definition is as follows:

```
class Date

instance variables

day    : Day;
month  : Month;
year   : Year;
inv if month in set {4,9,6,11}
    then day <= 30
    else (month = 2) => (day <= 29)

types

Day = nat1
```

```
inv d == d <= 31;

Month = nat1
inv m == m <= 12;

Year = int;

end Date
```

Recall the invariant on the instance variables presented in Section 4.3. Note how the use of separate type definitions for days and months allows their specific restrictions to be stated separately from the restrictions that span several instance variables. Under this definition, a value of type Day may never exceed 31, regardless of whether it is embedded in a Date object. When defining invariants, it is worth giving some consideration to its appropriate placement.

Suppose that the full CPR number, with the birth date and the four extra digits, is to be modelled by the class CPRNo. The details of this class will be considered later. For now, consider the class modelling a patient. A patient's attributes include their name, their CPR number and their history. An outline of part of the Patient class might be as follows:

```
class Patient

instance variables

   name : [seq of char];
   cprno : CPRNo;
   history : seq of Event

types

public
Sex = <Male> | <Female>;
Event = token;

end Patient
```

The sex of a patient is not modelled as a separate attribute, as this can be determined from the CPR number. An operation could be supplied to return the sex, so the type Sex is public. The details of events are not of particular interest in the model, so the modeller has chosen to use token as the representation for this type.

```
class CPRNo

instance variables

   date : Date;
   code: Digits4;

types

public
Digits4 ::
   d1 : Digit
   d2 : Digit
   d3 : Digit
   d4 : Digit;

Digit = nat
inv d == d <= 9

end CPRNo
```

This example models the special four-digit number as a public type and has an instance variable code corresponding to that component of the CPR number.

4.5 Values and Value Definitions

Values, like constants in programming languages, are used to give meaningful names to otherwise less meaningful values. If a value changes it need only be changed in one place, reducing the possibility of clerical error.

Values are defined in blocks within VDM++ class definitions. Each block starts with the keyword values, followed by an arbitrary number of value definitions separated by semicolons (;):

```
class class name

values
value definition 1;
value definition 2;
...
value definition n

end class name
```

where a value definition has the form:

```
value name : type = value
```

Note that if the *value* expression refers to another *value name* then from a VDMTools perspective the definition must be present prior to the usage. This is the only place where an ordering between the different member definitions makes a difference.

For example, a value representing the credit limit for a bank account may be expressed as follows:

```
CreditLimit: Euro = 5000
```

where, of course, `Euro` is the name of a suitable type. The *type* part of a value definition is optional, but it is recommended to include it.

4.6 Summary

The basic constructs for defining data using the VDM++ language have been introduced. *Classes* are the primary means of structuring a VDM++ model; the objects instantiated from them constitute an object-oriented system. The basic elements of a class description are the *blocks* of definitions covering data (*instance variables*, *types*, *values*) and functionality (*functions* and *operations*).

Instance variables model the internal state of an object. They have data types, may be initialised and can be restricted by invariants. New data types can be defined, built from the very abstract basic types such as *numeric* and *character* types, and type constructors that build structured types such as *product types* and *record types* with named fields. Primitive values include structureless *tokens* and *quote types*, often combined with the *union type* to give enumerations. All the data types have associated operators, some of which are *partial*.

New types, specific to the model, can be defined and invariant restrictions recorded for them.

5

Defining Functionality

The aim of this chapter is to introduce the mechanisms available for describing functionality in a VDM++ model. It lays essential groundwork for the following chapters on modelling with collections and relationships.

5.1 Introduction

In a VDM++ model, the computations performed on data are modelled as *functions* and *operations* located within objects. A function takes input parameters and produces a result, with no reference to the instance variables of the object. An operation also takes inputs and returns a result, but it may read and modify the instance variables. While functions *must* always return the same result for a given input, operations *may* return different results on each application because they may depend on the values in the instance variables at the time of calling.

The wide-spectrum character of the approach advocated in this book is reflected in the levels of abstraction that may be used in defining functionality. A function or operation definition can take one of two forms: *explicit* or *implicit*.

An explicit definition describes the result directly in terms of the inputs (and instance variables in the case of an operation). An implicit definition does not define a unique result but rather specifies the essential properties required of the result. Such definitions leave open the question of how the result is to be calculated. Although they are less readily executed, implicit definitions are much easier to reason about than complex explicit algorithmic definitions. The following sections introduce the explicit and implicit styles in more depth.

In a VDM++ model, function and operation definitions are defined in blocks in the usual way, following the keywords `operations` and `functions` respectively:

```
class class name

operations
operation 1;
operation 2;
...;
operation n

functions
function 1;
function 2;
...;
function n

end class name
```

Section 5.2 introduces function definitions in both explicit and implicit forms. Section 5.3 does the same for operations and addresses operation invocation and constructors, used to create objects and set their instance variables. Both function and operation definitions rely on the basic repertoire of expressions and statements that underpin the formal modelling language. These are described in Section 5.4 and 5.7. Throughout the chapter, reference will be made to the EPR example given in Chapter 4.

5.2 Function Definitions

A VDM++ function takes inputs and returns results that depend solely on the inputs and not on instance variables. It is important to remember that it has no side effects, in contrast with the "functions" found in many programming languages.

5.2.1 Implicit Function Definitions

Recall that VDM++ is a modelling language, not a programming language, and that models are built for the purpose of analysis, not exclusively for execution. Implicit definition permits description of the desired properties of the function without requiring algorithmic detail that may not be relevant to the analysis and may unfairly bias subsequent implementation and design.

An implicit function definition gives the properties required of the result, without necessarily giving an expression or algorithm for its computation. The properties required are expressed in a *postcondition*. A postcondition is a boolean expression that describes the relation between the inputs and result of the function. For example, the function sqrt returns the square root of its input:

```
class class name

functions

sqrt (x:nat) r: real
post r * r = x

end class name
```

where r is an identifier indicating the result.

The postcondition states the required property, that the returned result when squared yields the input. Notice also that the writer has left it open whether the positive or negative root is returned, perhaps because this does not matter for the model in question. A function returning the positive root could be defined by giving a stronger postcondition:

```
sqrt (x:nat) r: real
post r * r = x and r >=0
```

It would be difficult to give an explicit definition of this function without entering into the detail of, say, a recursive approximation algorithm. The implicit definition is, by comparison, clear and precise about *what* is to be computed.

Suppose the domain of the function were to be widened to, say, include rationals:

```
sqrt (x:rat) r: real
post r * r = x
```

This function definition cannot be satisfied, because the type rat includes negative numbers for which it is not possible to compute a result satisfying the postcondition. To record a restriction on the use of a function, a precondition is used. A *precondition* is a logical expression stating the conditions under which the modeller expects the function to be applied. The square root function could be redefined as follows:

```
sqrt (x:rat) r: real
pre x >= 0
post r * r = x
```

Misapplication of functions to inputs not satisfying the precondition is undefined in VDM++. The analyst of the model cannot conclude anything about what might happen under such circumstances. If the modeller wants to ensure that, for example, an error is returned on misapplication, this has to be defined explicitly, e.g.,

```
sqrt (x:rat) r: [real]
post if x >= 0
     then r * r = x
     else r = nil
```

Note that the return type has been made optional to accommodate the `nil` error value. In general an implicit function definition has the following syntax:

```
function name (parameters) result name and type
pre predicate
post predicate
```

The precondition may be omitted, in which case it defaults to `true`, meaning that the function is intended to be total (it can be applied to all inputs of the correct type). The *result name* and *type* are separated by a colon (":").

5.2.2 Explicit Function Definitions

An explicit function definition defines the types of the input and result and gives an expression that denotes the result in terms of formal parameters. For example, a function that returns the greater of two input values can be defined as follows:

```
max: int * int -> int
max(v1, v2) ==
   if v1 > v2
   then v1
   else v2
```

The syntax for an explicit function definition is as follows:

```
function name: parameter types -> result type
function name(parameters) ==
   expression
pre predicate
post predicate
```

This definition has the following components:

- The *signature* with the function name, the types of input parameters and the type of the result separated with a "->" symbol.
- The name of the function is repeated followed by a list of parameters. These are usually identifiers but they can also be patterns (see Section 5.5).
- The *function body* is an expression that, after substitution of the function's formal parameters with its actual parameters, evaluates to the result of the function. In the preceding example, an if-then-else expression is used to distinguish between the two values v1 and v2. This and other expressions that can be used for function bodies (and at many other places in the language) are introduced in more detail in Section 5.4.
- The *precondition*, as with implicit function preconditions, is a logical expression stating the conditions under which the function is to perform.
- If there is a *postcondition*, it is possible to refer to the result value using the keyword RESULT.

Functions are applied to actual parameters in the way familiar from programming. For example, the following expression

```
max(4,-3)
```

denotes the application of max to the specific input values 4 and -3 and would evaluate to 4. The syntax for function application is as follows:

```
function name(arguments)
```

where the arguments can be expressions of the types required in the function definition. It yields the result of evaluating *function name* against *arguments*.

An explicit function definition can also be recursive, calling itself either directly or indirectly through another function. A well-known example of this is the factorial function, which can be defined as follows:

```
fac: nat1 -> nat1
fac (n) ==
    if n > 1
    then n * fac(n-1)
    else 1
```

So fac(4) would yield the result 24 (4 * 3 * 2 * 1).

Functions in the EPR Example

Chapter 4 introduced the EPR example. Recall that in that model, people are identified by their CPR number and the uniqueness of numbers is maintained through a central population registry. Consider here a function to determine the sex of a person from the unique four-digit part of their CPR number. This would appear in the functions block of the class definition. It would take the four-digit number as input and return the sex of the patient. The signature would thus be:

```
Sex: CPRNo'Digits4 -> Patient'Sex
```

The function is quite straightforward and could be described explicitly as follows:

```
Sex: CPRNo'Digits4 -> Patient'Sex
Sex(code) ==
  if code.d4 mod 2 = 0
  then <Female>
  else <Male>;
```

The process of getting a new CPR number involves looking at the existing register (the id instance variable), so that must be represented as an operation rather than a function.

Suppose that a function is required to generate a sample of people from a population, given the size of the sample required and the sex that all members of the sample must have in common. This sort of function could be required for public health studies. The function for performing the selection might be modelled implicitly as follows to avoid giving a set of detailed criteria for selection:

```
Sample(pop: set of CPRNo, sexreqd:Patient'Sex, size:nat1)
       s:set of CPRNo
post card s = size and
     forall p in set s &
        p in set pop and Sex(p) = sexreqd
```

The operator card gets the size (cardinality) of a set. Observe that the Sex function can be used in this new function. This model is rather naïve: The function could not be satisfied if the population size is smaller than the sample. It would be appropriate to add a precondition:

```
Sample(pop: set of CPRNo, sexreqd:Patient'Sex, size:nat1)
       s:set of CPRNo
pre card pop >= size
post card s = size and
     forall p in set s &
        p in set pop and Sex(p) = sexreqd
```

Exercise 5.1 The `Sample` function might still not be satisfied if there are not enough people of the correct sex in the population to allow the sample to be supplied. How would you add a restriction to the function to indicate that all the members of the given population must be of the correct sex? Modify the function to return an error value if the population is not big enough to supply the sample. □

5.3 Operation Definitions

Operations take inputs and return results but may also read from and write to instance variables. As with functions, they may be defined implicitly or explicitly.

5.3.1 Implicit Operation Definitions

Implicit operation definitions use a postcondition in much the same way as implicit function definitions do. The difference is that, in addition to referring to inputs and result, an operation postcondition can refer to the "before" and "after" values of instance variables.

As a simple example, consider a class representing bank accounts and an operation to withdraw a sum of money from a bank account, returning the updated balance. Described implicitly, the operation would take the following form:

```
class class name

instance variables
balance: int
limit   : int

operations
Withdraw(amount: nat) newBalance: int
ext rd limit : int
    wr balance : int
pre balance - amount > limit
post balance + amount = balance~ and newBalance = balance

end class name
```

This *header* gives the operation name, the names and types of input parameters and the name and type of the result (if a value is returned by the operation). Input parameters and result combined, including their types, are referred to as the *operation signature*.

The precondition refers to the instance variables before the operation is invoked. The postcondition refers to the instance variables both before and after. The "before" values are decorated with a trailing tilde ("~").

The "`ext`" component, called the externals list, frames the instance variables that are involved in the implicitly defined operation. The keyword "`rd`" indicates read-only access, where "`wr`" indicates read and write access. Instance variables not mentioned in the externals list may not be accessed or changed by the operation.

The syntax of an implicit operation definition is as follows:

```
operation name(parameters) result name and type
ext instance variables read and write clauses
pre predicate
post predicate
```

5.3.2 Explicit Operation Definitions

In an explicit operation definition, the body consists not of an expression denoting the result but of a sequence of *statements* giving an algorithm for computing the result and changing the contents of the instance variables. VDM++ has a rich collection of statements, which are further introduced in Section 5.7.

Consider the `Withdraw` operations from the banking example. This could be defined explicitly as follows, using an assignment statement is used to assign a new value to the instance variable `balance`.

```
class class name

instance variables
balance: int
limit  : int

operations
Withdraw: nat ==> int
Withdraw(amount) ==
( balance := balance - amount;
  return balance
)
pre balance - amount > limit

end class name
```

An explicit operation definition has the following syntax:

```
operation name: parameter types ==> result type
operation name(parameters) ==
  statements
pre predicate
post predicate
```

The precondition and postcondition are optional. The postcondition has the same meaning as for implicit operation definitions (introduced later) and should be compliant with the operation body. It serves as an extra check on the correctness of the operation body. Just like with postconditions for explicit function definitions the result of the operation can be referred to using the RESULT keyword.

Operations in the EPR Example

Examples of different operation definition styles arise in the EPR example. First, consider an operation that allows the generation of a new unique four-digit component for a new CPR number. To ensure that the new code is unique, it is necessary to extract the existing codes from already allocated CPR numbers. An operation could be defined in the CPRNo class to return this unique code from a CPR number.

```
class CPRNo

instance variables

code: Digits4;

. . .
operations

public GetCode: () ==> Digits4
GetCode() ==
  return code;

end CPRNo
```

This simple explicit operation returns the value of the instance variable. Although the operation is very simple, it is still worth having. Among other things, it makes it easier to change the representation in the instance variable without having to modify the users of the class.

Now consider the central registry, modelled in the CPR class. There could be a single instance variable, modelling the collection of allocated CPR numbers. Let this be a mapping from birth date to CPR numbers: the Date class was defined in Chapter 4.

Mappings will be treated in detail in Chapter 8 and sets in Chapter 6. Here it is enough to know that mappings can be applied using a syntax identical to that of function application. The priority for the model is to ensure that the special four-digit component is indeed unique; how to go about generating it is not such a concern. This "what, not how" character immediately suggests the use of an implicit operation definition. A first attempt at such a definition might be as follows:

```
class CPR

instance variables

  id_m : map Date to set of CPRNo

operations

public
  GenerateUniqueCode(d:Date, sex: Patient'Sex)
                         dr:CPRNO'Digits4
  ext rd id_m : map Date to set of CPRNo
  post Sex(dr) = sex and
        forall cprno in set id_m(d) &
          cprno.GetCode() <> dr.GetCode()

end CPR
```

The postcondition asserts that the result must be of the right sex and that the code should not already be present in the registry for the given birth date. To the experienced modeller, this looks suspicious because there is no guarantee that the date in question is actually present in the registry. A precondition could remedy this deficiency:

```
class CPR
operations

public
  GenerateUniqueCode(d:Date, sex: Patient'Sex)
                         dr:CPRNO'Digits4
  ext rd id_m : map Date to set of CPRNo
  pre d in set dom id_m
  post Sex(dr) = sex and
        forall cprno in set id_m(d) &
          cprno.GetCode() <> dr.GetCode()

end CPR
```

Still there is no guarantee that there are actually any unallocated codes available for a given birth date. This would cause a problem if more than about 5000 children of the same sex were born on the same day. However, recording this additional restriction is beyond the scope of this introductory chapter.

5.3.3 Deferral, Delegation and Abstract Classes

Sometimes it is not possible or desirable to provide a complete definition of a function or operation. For example, at an early stage of development it may not be useful to provide a definition, or insufficient information may be available for a definition. In these situations it is possible to defer the definition to a later stage. A practical advantage of this is that the model can still be checked using VDMTools at intermediate stages prior to completion. Deferral is indicated by the keyword "is not yet specified." For example, a deferred function definition would have the following syntax:

```
function name: function signature
function name(parameters) ==
  is not yet specified
```

Within an inheritance hierarchy, it may not be possible to give a definition for a function or operation because it depends on additional information provided in a subclass. In such a situation it is possible to *delegate* the definition to the subclass. This is denoted by the keyword "is subclass responsibility". For example, a delegated function definition has the following syntax:

```
function name: function signature
function name(parameters) ==
  is subclass responsibility
```

A class containing a delegated definition is termed an *abstract class*. Instantiating an object from an abstract class is meaningless and therefore not allowed in VDM++. If the subclass does not provide a definition for the delegated operation then this subclass is also abstract. For example, an abstract base class for ordered collections might define a delegated sorting function; concrete subclasses would then provide specific algorithmic implementations of this function. Examples of using delegation are given in Chapters 7 and 9.

5.3.4 Constructors

Constructors are a special form of operation that have the same name as the class in which they are defined. They create a new instance of that class and return an object reference to that instance; their return type must therefore be that same class name. If a return value is specified then this should be self although this can be omitted.

Multiple constructors can be defined in a single class because in VDM++ operations can be *overloaded*. This simply means that the same name can be used for different operations as long as they are defined with different parameter types.

Now that operations are available to generate unique four-digit codes, the operation to build a new CPR number can be completed in the CPRNo class:

```
class CPRNo

instance variables

date : Date;

code: Digits4;

...
operations

CPRNo: Date * Patient`Sex * CPR ==> CPRNo
CPRNo(d,sex,cpr) ==
   (date := d;
    code := cpr.GenerateUniqueCode(d,sex)
    );

end CPRNo
```

5.3.5 Operation Invocation

After defining an operation, it is possible to invoke it. For example, consider an operation op defined in a class C:

```
class C

operations
op: nat ==> ()
op (n) ==
   ...

end C
```

and suppose there is an object instantiated from this class, referred to by an identifier c. Then op can, for example, be invoked as follows:

```
c.op(3)
```

Note that c must be of object reference type (see Section 4.4.2) class C.

Invocation of an operation is also referred to as a VDM++ call statement if it occurs where a statement is expected (see Section 5.7).

5.4 Expressions

Expressions are used to describe calculations that are free of side effects, meaning that an expression can never affect the value of an instance variable unless it contains a call to an operation. Expressions have already been illustrated in explicit function definitions and within preconditions, postconditions and invariants.

An expression can be evaluated by replacing the identifiers used in the expression with actual values. The evaluation of an expression results in a single value. VDM++ has about 25 different categories of expressions, not all of which are used equally often. The most fundamental ones are introduced in this section; others are gradually introduced in subsequent chapters alongside the development of different models. For a complete overview, see [LangManPP], which can also be obtained via the Web site that accompanies this book.

5.4.1 Let and Def Expressions

Let expressions and *def expressions* are used to improve the readability of complex expressions. They are used to define a name to stand for a more complex term.

A let expression has the form:

```
let pattern = expression1 in
   expression2
```

where *pattern = expression1* may repeatedly occur, separated by commas, and *pattern* may be associated with a type. The value of the entire let expression is the value of the expression *expression2* in the context in which *pattern* is matched against the expression *expression1*. Pattern matching will be treated in Section 5.5, but for now it is sufficient to consider simple identifiers. For example, the expression

```
let a = 2, b: nat = a+7 in
   a+b+15
```

will yield the value 26.

A def expression is written as follows:

```
def pattern = expression1 in
    expression2
```

where the same restrictions and possibilities as for the let expression hold. A def expression is typically used in operations; it is similar to a let expression except that *expression1* may contain references to instance variables (effectively meaning that the evaluation of a def expression depends on the values held in instance variables at the point of evaluation).

5.4.2 Conditional Expressions

VDM++ has two forms of conditional expressions: the *if-then-else expression* and the *cases expression*.
    The if-then-else expression has the form:

```
if predicate
then expression1
else expression2
```

The value of the if-then-else expression is the value of *expression1* evaluated in the given context if *predicate* evaluates to true. Otherwise the value of the if expression is the value of *expression2* evaluated in the given context.
    The cases expression has the form:

```
cases expression:
  pattern list 1 -> expression 1,
  pattern list 2 -> expression 2,
  ...            -> ...,
  pattern list n -> expression n,
  others         -> expression others
end
```

where *expression* is an expression of any type. The *pattern list* is matched one by one against the expression *expression*. The *expression* is are expressions of any type, and the keyword others and the corresponding expression *expression others* are optional. The value of the cases expression is the value of the *expression x* expression evaluated in the context in which one of the *pattern list x* patterns has been matched against *expression*. The chosen *expression x* is the first entry where it has been possible to match the expression *expression* against one of the patterns. If none of the patterns matches *expression* an others clause must be present, and then the value of the cases expression is the value of *expression others* evaluated in the given context.

5.4.3 Quantified Expressions

Quantified expressions are a kind of logical expression used frequently when it is necessary to make an assertion about a collection of values. There are two forms: *universal quantification* and *existential quantification*. Each binds one or more variables to a type or a value from a set and evaluates to a boolean value. Both forms of quantified expression were seen in the alarm example in Chapter 2.

Universal quantification has the form:

```
forall bind list & predicate
```

The expression evaluates to the boolean value `true` if *predicate* evaluates to `true` *for all* pattern identifiers introduced in *bind list* and to false otherwise. In Section 5.2.2 we saw an example of this kind of universal quantification.

Existential quantification has the form:

```
exists bind list & predicate
```

The expression evaluates to the boolean value `true` if *predicate* evaluates to `true` *for at least one* pattern identifier introduced in *bind list* and to false otherwise. In Chapter 6 a number of examples of existential quantification are presented.

5.4.4 Record Expressions

The reader will by now be familiar with the use of record constructors (mk) to build record values from their component parts. There is a further, very useful, expression that can be used to produce a record value, based on the value of another record. This is the *mu expression* (mu is short for "mutation"). It has the form:

```
mu (record expression, record modifiers)
```

where *record expression* evaluates to a record, and each *record modifier* has the form:

```
field identifier |-> expression
```

The result of the mu expression is a record identical to *record expression*, except that the value of each field identified by a *field identifier* is replaced by the evaluation of its corresponding *expression*. The mu expression is particularly useful when defining functions or operations that change just one or two fields of large record structures because it eliminates the need to repeat the values of all the unchanged fields. The mu expression can be seen in use in Chapter 12.

Exercise 5.2 What would the following expressions yield as results? Remember to evaluate these by hand before using the tools.

1. ```
   let x1 = 3,
       x2 = 6.5
   in let x3 = x1 + x2
      in x3 * x1
   ```
2. ```
   let x1 = 3,
       x2 = 6.5,
       x3 = x1 + x2
   in x3 * x1
   ```
3. ```
   let x = 2
   in
      if x mod 2 = 0
      then <Even>
      else <Odd>
   ```
4. ```
   let sign = <Unknown>
   in cases sign:
      <Even> -> "Even",
      <Odd>  -> "Odd",
      others -> "Unknown"
      end
   ```
5. Assuming the type

   ```
   Coord::x:nat
         y:nat
   ```

   then the expression

   ```
   let c1 = mk_Coord(0,0),
       c2 = mk_Coord(5,7)
   in mu(c1, x |-> c2.x)
   ```

□

5.5 Pattern Binding

VDM++ includes a notion of pattern matching that proves very useful in situations where an identifier is to be bound to a concrete value. Patterns themselves are like empty shells that have a meaning only once they are matched against a concrete value. The result of such pattern matching is that the pattern identifiers from the pattern are bound to concrete values. These concrete bindings can then form a context in which a calculation can be carried out. Examples of such binding have already been seen. The example of a simple let expression shows two simple pattern matches of identifiers to values:

```
let a = 2, b: nat = a+7 in
    a+b+15
```

In addition to an identifier a few other types of patterns are used in this book. They are:

Match Values: In traditional programming languages the alternatives in a cases expression are constant values. In VDM++ they can be general patterns, but to cater for the more traditional cases expression a pattern in VDM++ simply can be a literal value, e.g., 7 or `true`.

Don't Care Patterns: This type of pattern is used to indicate to the reader that the actual value matched against this type of pattern does not matter for the subsequent calculation. It is written as a dash "-" and is mainly used in parameter lists for functions and operations that are deferred to subclasses.

Record Patterns: This type of pattern can be used to match against record values. It is written in the same way as the record expressions shown in Section 4.4.2 except there are patterns instead of values inside the mk constructor. It takes the following form:

```
mk_tag (pattern 1, pattern 2, ..., pattern n)
```

This record pattern will only match a record value that has the same *tag* and the same number of fields, and each of the component values matches against the corresponding pattern. This kind of pattern was used in Section 4.4.4.

There are more types of patterns available in VDM++ than can be discussed here. For a complete overview see the Language manual [LangManPP], available via the Web site accompanying this book.

5.6 Bindings

Pattern matching allows concrete values to be bound to identifiers that are used in subsequent calculations. Where the values range over a whole collection of possibilities such as a type or a set, a binding is used. VDM++ differentiates between *set bindings* and *type bindings*. The syntax for a set binding is:

```
pattern in set set expression
```

and the syntax for a type binding is:

```
pattern : type
```

For example, in the `Sample` function, the postcondition included the following expression with a set binding:

```
forall p in set s & p in set pop and Sex(p) = sexreqd
```

Here the variable p is a simple identifier pattern to be bound to each value in the set s and the rest of the expression evaluated for each binding.

The distinction between the two types of binding is significant because, as has already been pointed out, sets are finite but types are unconstrained. A consequence is that type bindings cannot be executed using the interpreter from VDMTools. Imagine a tool evaluating the expression

```
forall x in set {1,3,5,7} & x**2 < 40
```

It is likely that a tool will have a routine that loops through all the possible bindings of a `forall` expression, evaluating the body of the `forall` for each binding. In this case, the loop will iterate through the possible bindings of x, evaluating the Boolean x**2 < 40 and returning `false` because 7**2 is 49. Now imagine the same mechanism evaluating the expression

```
forall x : nat & x**2 < 40
```

Without some special knowledge about the theory of numbers, which would only be useful for evaluating this expression, the loop will never terminate. These are the sorts of conditions under which validation by techniques such as proof and inspection are needed to cope with the level of abstraction in the model.

Both simple identifiers and more sophisticated patterns can be used in bindings. They are typically used in quantified expressions (See Section 5.4.3) and the *comprehension* expressions used to generate collections. Comprehensions are introduced in Chapters 6 to 8.

5.7 Statements

Statements are used in the bodies of explicit operations to describe an algorithm, typically affecting the instance variables in a class. A statement can be executed, as a result of which the value of one or more variables is changed. The most fundamental kinds of statement are introduced in this section; others are gradually introduced in subsequent chapters alongside the development of different models. For a complete overview see the language manual [LangManPP], available via the Web site accompanying this book.

5.7.1 Let, Def and Block Statements

In much the same way VDM++ has let expressions and def expressions (see Section 5.4.1), the language also has *let statements* and *def statements*. The difference is that, instead of introducing identifiers that are used inside expressions, here they are used in statement bodies. Let statements and def statements are therefore written as follows:

```
let pattern = expression in statement
```

and

```
def pattern = expression in statement
```

Statements that are intended to be executed sequentially can be gathered into a *block statement*. As with the *let* and *def* statements, new identifiers can be introduced and used in the body of the block statement. However, whereas let and def statements introduce new identifiers essentially representing values (identifier *definition*), the identifiers introduced in block statements represent declarations of *local variables* whose value can be changed in the statement body. A block statement is written as follows:

```
(dcl declarations;
  statements
)
```

where the "dcl *declarations*" (optional) represents a number of declarations separated by commas, each of which has the following form:

```
identifier : type := expression
```

Each declaration introduces a local variable with name *identifier* and of type *type*, with an initial value defined by *expression*. The ":= *expression*" part is optional. The body of the block consists of statements, separated by semicolons, in which the declared local variables can be used. These are executed sequentially.

5.7.2 The Assignment Statement

The assignment statement changes the value of an instance or local variable. Its syntax is as follows:

```
identifier := expression
```

The effect of an assignment is that the *expression* is evaluated and the result is assigned to the *identifier*. The *identifier* must either be an instance variable or a locally declared identifier in a block statement. For example, the following VDM++ statement creates local variables and assigns a value to them:

```
let x = 2
in
( dcl signStr : seq of char,
      y : nat := x;
  signStr := if y mod 2 = 0
             then "Even"
             else "Odd"
)
```

In its most general form, the left-hand side of the assignment statement can be a *state designator* indicating the part of the state that should be modified. This is illustrated in Chapter 8.

5.7.3 Conditional Statements

VDM++ has two forms of conditional statements: the *if-then-else statement* and the *cases statement*.

The conditionals are identical to their expression counterparts (see Section 5.4.2), except that the evaluation of a condition leads to execution of a statement rather than evaluation of an expression. In addition, the else part of an *if-then-else statement* is optional (if absent, it corresponds to an else skip statement) and, similarly, the *others* part of a cases statement is optional.

5.7.4 Loop Statements

Two kinds of loop statements can be distinguished: *for loop statements* and *while loop statements*. A for loop statement has the form:

```
for identifier = expression1 to expression2 do
    statement
```

The for loop is executed by assigning *identifier* the evaluation of *expression1*, executing *statement* (in which *identifier* may occur), incrementing *identifier* by 1 and repeating the process until *identifier* has reached the value of *expression2*. Variants of the for loop statement are described in Chapters 6 and 7.

A while loop statement has the form:

```
while predicate do
    statement
```

where *predicate* is a boolean expression that is evaluated before *statement* is executed. If the predicate returns `true` then *statement* is executed, after which the predicate is evaluated again. Execution of *statement* stops when the predicate no longer holds, i.e., it returns the value `false` or the *statement* in the body of the loop returns a value.

5.7.5 The Skip and Return Statements

In VDM++ there is a simple statement called `skip`, which does nothing. This is typically used when an operation or a part of an operation has to perform a null action. Chapter 9 includes examples of its use in situations where subclasses redefine an operation.

The return statement is used to terminate execution of an operation and to return a specific return value to the environment from which the operation was called.

A return statement has the form:

```
return expression
```

This was shown in Section 5.3.2.

Note that the constructor of a class has a built-in (implicit) "`return self`" statement at the end as explained in Section 5.3.4.

Exercise 5.3 Define an operation `fac: nat ==> nat` that uses a while loop to compute the factorial of the natural number parameter to the operation. This factorial is returned as the result of the operation. □

5.8 Summary

There are several ways in which functionality can be expressed in VDM++ models, each approach providing different abstraction capabilities. *Functions* make no reference to instance variables. In contrast, *operations* can modify the values of instance variables and return results. Both functions and operations can be modelled *implicitly* or *explicitly*. An implicit approach to implicit function definition defines the relation between the results and inputs by means of a *postcondition*. In the case of an implicitly specified operation, the postcondition is extended to include the instance variables before and after execution. An explicit definition of a function gives an expression denoting the result. An explicit operation definition gives a program-like series of statements describing the computation and the effects on instance variables. All function and operation definitions can be restricted by *preconditions*, recording the assumptions about inputs and instance variables when the function or operation is applied.

More advanced kinds of class members such as threads and synchronisation constraints are deferred until Chapter 12. This chapter ends with an introduction to expressions, pattern matching, bindings and statements in VDM++.

6

Modelling Unordered Collections

This chapter aims to introduce the set type constructor and demonstrate its use modelling collections of data that have no inherent order. On completion of this chapter the reader should understand the defining properties of sets, understand and be able to use the basic set operators and understand how sets can be used to decompose and simplify a large model.

6.1 Introduction

Sets are the among most fundamental means of data abstraction. A *set* is a collection of values in which order and repetition are not significant. This is a long way from the traditional view of data stored in computer memory as a linear sequence of bytes in a particular order and with repetition. Of course, when the time comes to implement a model, some commitment to an order for the collection must be made. However, at the modelling stage, there are significant benefits to be gained from using sets. These benefits will be demonstrated in this chapter.

A large part of this chapter is devoted to an example that is used to demonstrate the use of sets as an abstraction mechanism when developing nontrivial models. The example used is that of a robot controller, an adaptation of a larger model.

6.2 Overview of Sets

A set is a finite collection of data. Set values are represented using "{" and "}" as delimiters, with individual values separated by commas. For instance, the value

```
{1, 2, 3}
```

is a set that contains three elements: namely the natural numbers 1, 2 and 3. This set value therefore has the type `set of nat`. In general, a set type in VDM++ can be written as

```
set of Type
```

which indicates all finite sets with elements of type *Type*, including the set with no elements at all.

Sets have two distinguishing properties: They are *unordered* and *repetition is insignificant*. What do these terms mean? *Unordered* means that the order in which the values in a set appear is not significant. Thus the following three sets are all equal:

```
{1, 2, 3}
{1, 3, 2}
{2, 3, 1}
```

Repetition is insignificant means that the set value is the same if it contains just one instance of a value or several instances of that value. For example, the following sets are all equal:

```
{1, 1, 2, 3}
{1, 1, 2, 2, 3, 3}
{1, 2, 1, 2, 3, 3, 1, 2}
```

Exercise 6.1 For each of the following pairs of sets, which pairs are equal?

1. $\{1, 3, 2\}$ and $\{3, 3, 1, 2, 1, 3\}$
2. $\{3, 1, 2, 1\}$ and $\{1, 2\}$
3. $\{$<Apple>, <Orange>, <Orange>$\}$ and
   $\{$<Orange>, <Pineapple>$\}$

□

A value contained in a set is known as a *member* of the set. The preceding expressions, in which the set members are enumerated one by one, are known as *set enumeration* expressions. It is possible to check whether a value is a member of a set using a *set membership* expression. Given a value v and a set s, this would be written as:

```
v in set s
```

This is a boolean expression that yields the value `true` if v is a member of s and `false` otherwise. For example, the expression

```
2 in set {1, 2, 3, 1}
```

yields the value `true` whereas the expression

```
2 in set {1, 3, 1}
```

yields the value `false`.

The set that contains no members is known as *the empty set* `{ }`. Testing membership of the empty set always yields `false`. A set containing just one member is called a *singleton set*.

Exercise 6.2 Calculate the value of the following set membership expressions:

1. `3 in set {1, 2, 3, 3, 1}`
2. `<Apple> in set {<Orange>, <Pineapple>}`

□

Sets of simple values are common, but more sophisticated structured models often call for sets that contain sets as members. For example `{{3, 5}, {}}` is a set that contains the sets `{3, 5}` and `{}`. Sets can also contain elements of apparently diverse types. For example, the set `{9,{3}}` contains the values 9 and `{3}`. This is well typed because all the elements belong to the union type `nat | (set of nat)`. The expression is itself of type `set of (nat | (set of nat))`. Thus the expression

```
{} in set {3, 5, {}}
```

yields the value `true`.

Exercise 6.3 Calculate the value of the following set membership expressions:

1. `{} in set {{1, 2}, {}, {3, 4, 5}}`
2. `{} in set {}`

□

Set enumerations are convenient for describing unordered collections with just a few members, but not for anything larger. Moreover, it is often desirable to be able to define a set based on a common property shared by the members of this set rather than just listing them. This is essential if we are to be able to describe sets that are formed dynamically from other collections, rather than being known statically at design time. For these situations, *set comprehension* provides a valuable way to describe collections. A set comprehension expression has the following form:

```
{create a value using x | x in set s & test x}
```

Such an expression builds up a set by performing the following steps for each member of the set s:

1. Take a member x from s,
2. Evaluate the test,
3. If the test evaluated to true, create a value using x and put it in the resulting set.

For example, consider the set comprehension expression:

```
{x * 3 | x in set {1, 2, 3, 4, 5} & x mod 2 <> 0}
```

This builds a set value in the following way: Any member of $\{1, 2, 3, 4, 5\}$ is taken, say the value 5. This value is then tested: 5 mod 2 is 1 so the test yields true. Thus the value 5 * 3 is added to the result. Next another member is taken from $\{1, 2, 3, 4, 5\}$, say 2. Again, this value is tested: 2 mod 2 is 0 so the test yields false and nothing is done with the member 2. Similar steps are performed for the other members of $\{1, 2, 3, 4, 5\}$. The result of the set comprehension expression is the set:

```
{3, 9, 15}
```

In fact, the test part of a set comprehension expression is optional. If absent, the test is regarded as true. For example, the expression:

```
{x * 3 | x in set {1, 2, 3, 4, 5}}
```

yields the set $\{3, 6, 9, 12, 15\}$. Type bindings can also be used in set comprehensions. Thus, it is possible to write:

```
{x | x: nat & x < 10}
```

which yields the set:

```
{0, 1, 2, 3, 4, 5, 6, 7, 8, 9}
```

Set comprehension expressions with type bindings cannot be executed by the VDM-Tools interpreter because that would involve evaluating the test for every member of a potentially infinite data type. Recall also that sets are finite structures, so care must be taken to ensure that the test is only satisfied in a finite number of cases.

Table 6.1: Summary of VDM++ set operators

Operator	Name	Type
e in set s1	Membership	A * set of A → bool
e not in set s1	Not membership	A * set of A → bool
s1 union s2	Union	set of A * set of A → set of A
s1 inter s2	Intersection	set of A * set of A → set of A
s1 \ s2	Difference	set of A * set of A → set of A
s1 subset s2	Subset	set of A * set of A → bool
s1 psubset s2	Proper subset	set of A * set of A → bool
s1 = s2	Equality	set of A * set of A → bool
s1 <> s2	Inequality	set of A * set of A → bool
card s1	Cardinality	set of A → nat
dunion ss	Distributed union	set of set of A → set of A
dinter ss	Distributed intersection	set of set of A → set of A
power s1	Finite power set	set of A → set of set of A

Exercise 6.4 Calculate the results of the following set comprehension expressions:

1. {y * 2 | y in set {1, 2, 3, 4, 5} & y mod 2 = 0}
2. {{i, i+1, i+2} | i in set {10, 11, 12}}
3. {y div 2 | y in set {1, 2, 4, 6, 9, 10}
   & y mod 2 <> 0}

□

The empty set, set enumerations and set comprehensions are the fundamental building blocks of models of unordered collections. VDM++ provides 13 operators for manipulating sets, summarised in Table 6.1. The use of sets and of these operators is illustrated through a substantial example in Section 6.3.

6.3 The Robot Controller

This section presents an example that demonstrates the use of sets and the operators listed in Table 6.1. The example is a system for navigating a robot from a start point, via a collection of waypoints to a final destination, where it performs some task, e.g., delivering a payload. An overview of the robot's motion is shown in Figure 6.1.

This robot maneuvers around obstacles that it is able to detect until it reaches a specified destination. When it arrives at its destination it completes its task (performs some action, delivers some payload, etc.) Its path from start point to destination point is defined by a collection of predefined navigation waypoints – coordinates describing a location in 2-D space. However the route from one waypoint to the next is determined dynamically by the robot, on the basis of the position of the next waypoint in relation to the last one reached and any obstacles detected. The robot is able to detect obstacles within a limited range, and therefore this range also limits the search space when determining a route.

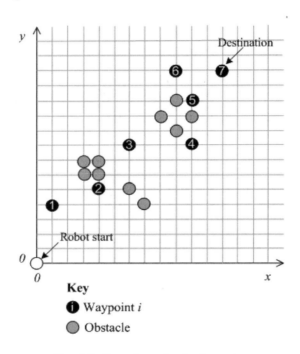

Fig. 6.1: Overview of robot's motion

The controller communicates with a number of subsystems:

Position Sensor: This is used to find the robot's current location and the direction in which it is moving.
Steering Controller: This controls the direction in which the robot travels.
Steering Monitor: A system used to ensure that the steering controller is operating within known safe boundaries.

The requirements for the controller are:

R1. The robot's current position is always available to the controller from a position sensor.

R2. The robot has a predetermined journey plan based on a collection of waypoints.

R3. The robot must navigate from waypoint to waypoint without missing any.

R4. The robot moves only horizontally or vertically in the Cartesian plane. It is not physically capable of changing direction with an angle greater than 90°. Attempts to do so should be logged.

R5. If the robot is off-course, i.e., it cannot find a route to the next waypoint, it should stop in its current position.

R6. The robot is able to detect obstacles in its path.

The main class in the model is `Controller`; this keeps track of the current position, computes the direction in which to travel and computes commands to change direction. It communicates with the subsystems listed earlier.

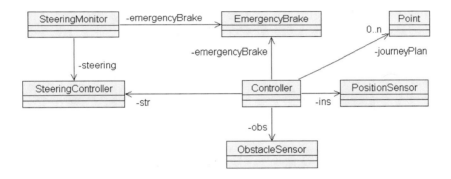

Fig. 6.2: Class diagram for robot controller

A UML class diagram representing the overall architecture of the controller is shown in Figure 6.2.

6.3.1 Points and Routes

Before describing the actual behaviour of the robot, it is necessary to describe how physical locations are represented in the model. These are modelled using the classes `Point` and `Route` (see Figure 6.3).

The class `Point` is used to represent physical locations; an instance of `Point` may be used to represent either a location in the robot's journey plan or the location of an obstacle detected by the robot. `Point` uses two instance variables to represent the 2-D Cartesian coordinates of the location to which the `Point` instance refers:

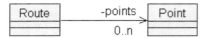

Fig. 6.3: The Route and Point classes

```
class Point
...
instance variables

x: nat;
y: nat;
end Point
```

The class Controller has an instance variable – journeyPlan – that is a set representing the predefined waypoints that this robot must visit:

```
class Controller
...
instance variables

journeyPlan     : set of Point;
end Controller
```

The instance variable index is used to represent the index in the journey plan:

```
class Point
...
instance variables

index: nat;
end Point
```

In reality an ordering between members would not be modelled in this way but because this chapter aims to introduce the first of the structured modelling techniques in VDM++ it is done using an index in the example presented here. In Chapter 7 this example will be revisited as an exercise for the reader, and it will then be clear how an ordering is most naturally modelled using VDM++.

The class provides a constructor that initialises these three instance variables:

```
class Point
...
public Point: nat * nat * nat ==> Point
Point(nx, ny, nindex) ==
( x      := nx;
  y      := ny;
  index := nindex;
);
end Point
```

For example, the journey plan shown in Figure 6.1 could be modelled using set enumeration:.

```
{new Point(1, 4, 1),   new Point(4, 5, 2),
 new Point(6, 8, 3),   new Point(10, 8, 4),
 new Point(9, 11, 5),  new Point(8, 13, 6),
 new Point(11, 13, 7)}
```

The class Route complements Point in that it is concerned with sets of Points. For instance, the static function GetPointAtIndex is used to locate a Point from a set of Points that has a particular index. How would this function be defined? First, an accessor operation for the instance variable index of class Point is clearly needed:

```
class Point
...
operations

public GetIndex: () ==> nat
GetIndex() ==
  return index;
end Point
```

The function GetPointAtIndex will then look something like:

```
public static GetPointAtIndex: set of Point * nat ->
                                        Point
GetPointAtIndex(pts, index) ==
  find that value p in the set pts
  where p.GetIndex() equals index
```

How can this function body be defined? For precisely this kind of situation, VDM++ provides an expression called *let be st* (pronounced "let be such that"). This expression takes the form

```
let x in set s be st predicate on x
in
    expression using x
```

The meaning of this is that a value x satisfying the given predicate will be found from the set s and then used to evaluate the *body* – the expression using x. If there is more than one member of s satisfying the predicate, a value is nondeterministically chosen; if no such value exists, the expression is not well-formed – in VDMTools this would lead to a run-time error. Typically a precondition is used to indicate to the caller that the *let be st* expression must be satisfiable.

In this case, the required predicate is:

```
p.GetIndex() = index
```

and for the body, the Point p is all that is needed. Adding a precondition to ensure that the expression is satisfiable gives the function:

```
class Route
...
functions

public static GetPointAtIndex: set of Point * nat ->
                                        Point
GetPointAtIndex(pts, index) ==
  let p in set pts be st p.GetIndex() = index
  in
      p
pre exists p in set pts & p.GetIndex() = index;
end Route
```

Observe that the *let be st* expression is highly abstract; the property required of the value to be found is described without any reference as to *how* the value is to be located. Thus, the precondition is required to ensure that at least one value exists.

An alternative, more algorithmic, approach would be to use a *set for loop* statement. This has the following form:

```
for all x in set s do
    statement using x
```

This could loop through the set pts until an object with the suitable index is found:

```
class Route
...
  operations

  public static GetPointAtIndex: set of Point * nat ==>
                                  Point
  GetPointAtIndex(pts, index) ==
    for all p in set pts do
      if p.GetIndex() = index
      then return p
  pre exists p in set pts & p.GetIndex() = index

end Route
```

6.3.2 Using Sets of Points

Care must be taken when dealing with sets of object references. For example, consider the following expression:

```
new Point(1,1,1) in set {new Point(1,1,1)}
```

It would be tempting to assume that this expression yields the value true; in fact it yields false! This is because set membership is based on equality; that is x in set s yields true if there is some value x' in s such that x = x'. However, as described in Chapter 4 a property of object references in VDM++ is that for objects o1 and o2, o1 = o2 if and only if o1 and o2 are both references to the same object.

Exercise 6.5 Calculate the results of the following set membership expressions:

```
1. new Point(1,2,3) in set {new Point(1,1,2),
                            new Point(1,2,3),
                            new Point(0,1,1)}
```

2. let p = new Point(1,2,3)
   in p in set {p}
3. let p = new Point(1,2,3)
   in p in set {new Point(1,2,3)}

□

How then is it possible to test, for example, whether the current waypoint wp occurs in a set of obstacles (represented by Point) obs detected by the robot? The real question here is does there occur in obs any point that has the same coordinates as wp? The issue therefore is to try and use a membership test on a set of coordinates instead of on a set of object references. To allow this, the operation GetCoord is defined in Point, which extracts the pair representing the coordinates of a Point:

```
class Point
...
public GetCoord: () ==> nat * nat
GetCoord() ==
  return mk_(x,y);
end Point
```

The required set membership expression then involves converting the Point and set of Points to a pair and set of pairs respectively:

```
wp.GetCoord() in set {o.GetCoord() | o in set obs}
```

This method of checking set membership is used frequently both in this example and more generally when sets of objects are used.

6.3.3 Arriving at a Waypoint

Suppose now that the robot with journeyPlan as shown in Figure 6.1 reaches the first waypoint. The remaining journeyPlan corresponds to the set of Points:

```
{new Point(4, 5, 2),   new Point(6, 8, 3),
 new Point(10, 8, 4),  new Point(9, 11, 5),
 new Point(8, 13, 6),  new Point(11, 13, 7)}
```

However, it greatly simplifies the algorithm if the journeyPlan always satisfies three properties:

1. The next waypoint to be visited always has index 1.

2. The index of the final waypoint to be visited (i.e., the robot's final destination) always has index corresponding to the number of waypoints remaining to be visited.

3. The indices for intermediate waypoints have the expected value so, for example, the waypoint after the next waypoint has index 2 and so on.

Because the algorithm expects that journeyPlan always satisfies these properties, they should be expressed as an invariant on the instance variable journeyPlan. By looking at the example journeyPlan, it is not difficult to see that the three properties can be collapsed into one: The indices for the journeyPlan should correspond to the set of natural numbers with members 1, 2 and so on up to n, where n is the number of waypoints in the journeyPlan.

This gives two subproblems to solve: First, how is the number of waypoints in journeyPlan obtained? VDM++ provides the operator card , from the term cardinality, which yields the number of members in a set. For example,

```
card {1, 2, 3, 3}
```

yields 3 (recall that repetition is not significant for sets) and the following expression

```
card {}
```

yields 0.

The second subproblem concerns how to construct the set of all integers between 1 and card journeyPlan. For this, VDM++ provides the *set range* expression. This has the form

```
{lb, ..., ub}
```

where *lb* and *ub* are numeric expressions. It yields the set of integers less than or equal to *lb* and greater than or equal to *ub*. If lb > ub the expression yields the empty set.

The solution to the two subproblems is therefore the expression

```
{1, ..., card journeyPlan}
```

A simple set comprehension expression can be used to extract the indices from the instance variable journeyPlan. The desired invariant is therefore:

```
class Controller
...
instance variables

inv {p.GetIndex() | p in set journeyPlan} =
    {1,..., card journeyPlan};
end Controller
```

In this case the number of waypoints in journeyPlan is therefore given by

```
card journeyPlan
```

Thus, from the initial journeyPlan, when the robot arrives at Point with x = 1 and y = 4, the journeyPlan should become the set

```
{new Point(4, 5, 1),  new Point(6, 8, 2),
 new Point(10, 8, 3), new Point(9, 13, 4),
 new Point(12, 13, 5)}
```

The operation to update the individual points of a journeyPlan (that is, each member of the set) in this way is straightforward:

```
class Point
...
operations

public TakeStep: () ==> Point
TakeStep() ==
( index := index - 1;
  return self
)
pre index > 1;
end Point
```

If the function to modify journeyPlan in this way is called TakeStep, the function will look like:

```
class Route
. . .
functions

public TakeStep: set of Point -> set of Point
TakeStep(pts) ==
  let laterPoints = those points in pts with index <> 1
  in
    {p.TakeStep() | p in set laterPoints};

end Route
```

Exercise 6.6 Complete the definition of TakeStep using set comprehension to model those points in pts with index <> 1. □

An alternative formulation of TakeStep would be to use the *set difference* operator. This takes the following form:

```
s1 \ s2
```

This expression yields all those members of s1 that are not in s2. For example,

```
{1, 1, 2, 3} \ {1, 3}
```

yields the set {2}. Also

```
{1, 2, 3} \ {-1, 4}
```

yields the set {1, 2, 3}. Observe that the members of the set on the right-hand side that are not present on the left-hand side of the operator have no effect on the value of the expression.

TakeStep could have been formulated as:

```
class Route
. . .
functions

public TakeStep: set of Point -> set of Point
TakeStep(pts) ==
  let firstPoint = GetPointAtIndex(pts, 1)
```

```
  in
     {p.TakeStep() | p in set pts \ {firstPoint}};

end Route
```

6.3.4 Controlling the Robot

As stated previously, the main system class is `Controller` and its instance variable `journeyPlan` is initially the predefined route, represented as a set of `Point`:

```
class Controller
...
instance variables

journeyPlan      : set of Point;
end Controller
```

`Controller` interacts with four different hardware devices: It takes input from `PositionSensor` and `ObstacleSensor` and generates commands that are sent via `SteeringController`; alternatively it is able to send commands to `EmergencyBrake` if, for instance, the robot is lost.

In the current context, modelling of these hardware devices is not the focus of interest. Therefore these hardware devices are represented as classes whose behaviour is not defined in this model. `Controller` has instance variables representing references to these devices, as shown in Figure 6.2:

```
class Controller
...
instance variables

obs              : ObstacleSensor;
emergencyBrake   : EmergencyBrake;
ins              : PositionSensor;
str              : SteeringController;
end Controller
```

The core of the controller is modelled by the operation `Update` defined in the `Controller` class. This operation repeatedly checks the current position and location of the robot and adjusts the steering so that the robot will reach the next waypoint. Thus a first attempt at this algorithm would look something like:

1. Find out the robot's current position.

2. Find out the next waypoint that the robot must visit.

3. If this waypoint has the same location as the current position then there are two possibilities:

Either this is the last waypoint, i.e., the robot has reached its final destination and can therefore complete its journey

or there are further waypoints to visit, in which case the journey plan must be updated.

Otherwise do nothing.

4. Calculate the commands needed by the steering controller to get the robot to this next waypoint.

5. Give these commands to the steering controller.

How is the robot's current position ascertained? From requirement **R1** it is known that the robot's current position is always available from the `PositionSensor`. Thus the operation call

```
ins.GetPosition()
```

can be used to ascertain the robot's current position. Now suppose that the identifier `currentPosition` is used to represent the result of this call.

The next waypoint that the robot is to visit must be obtained. From the invariant on `journeyPlan` described in Section 6.3.3 it is known that the next waypoint to be visited is the waypoint with index 1. Using the function `GetPointAtIndex` in the `Route` class described in Section 6.3.1, the following expression yields the next waypoint:

```
Route`GetPointAtIndex(journeyPlan, 1)
```

Having obtained the next waypoint, this must be checked against the current position. Recall from the discussion in Section 6.3.2 that simple comparison of objects would not give the expected result; instead coordinates need to be extracted. This gives a test of the form

```
Route`GetPointAtIndex(journeyPlan, 1).GetCoord() =
currentPosition.GetCoord()
```

If this test yields `true`, a further test is necessary: Is this the last waypoint? If this is the last waypoint, then the number of waypoints left in `journeyPlan` must be 1 (i.e., just the waypoint that has now been reached). This can be easily expressed using the `card` operator:

```
card journeyPlan = 1
```

If this is the case, the journey should be completed and the system stopped. The details of journey completion are not important at the moment, so an operation, named `CompleteJourney` is called to perform this task. Including the test, this gives something like:

```
if card journeyPlan = 1
then CompleteJourney()
else ...
```

If this test yields false, then the waypoint at which the robot has just arrived needs to be removed from the `journeyPlan`. The function `Route'TakeStep` introduced in Section 6.3.3 accomplishes this task:

```
journeyPlan := Route'TakeStep(journeyPlan)
```

Having updated the `journeyPlan`, the next step is to calculate the commands needed to get the robot to the next waypoint. This can be subdivided into two smaller steps:

4(a). Calculate a route from the current position to the waypoint, given the obstacles that the robot is currently able to detect. This is achieved using the operation `PlotCourse` described in Section 6.3.6. If no valid route can be found, this operation returns `nil`; in this case, following Requirement **R5**, the `Emergen-cyBrake` is enabled, causing the whole system to stop.

4(b). Convert this route into a command for the steering controller. This is performed using a function called `ComputeDesiredSteerPosition`.

Finally, the computed steering command is issued to the steering controller. This is performed by the operation `AdjustSteering`. The definitions of the operations `ComputeDesiredPosition` and `AdjustSteering` are straightforward; the details are not given here.

Putting these steps together gives the following operation definition:

```
class Controller
...
operations

Update: () ==> ()
Update() ==
let currentPosition = ins.GetPosition()
in
( if Route'GetPointAtIndex(journeyPlan,1).GetCoord() =
     currentPosition.GetCoord()
  then
```

```
        if card journeyPlan = 1
        then CompleteJourney()
        else
        ( journeyPlan := Route'TakeStep(journeyPlan);
          let obstacles = obs.GetData(),
              route     = PlotCourse(obstacles)
          in
            if route = nil
            then emergencyBrake.Enable()
            else
              def dfps = ComputeDesiredSteerPosition(
                            ins.GetDirection(),
                            route.GetPoint(2),
                            str.GetPosition())
              in AdjustSteering(dfps)
        )
);
end Controller
```

The critical stage here is the call to PlotCourse. It is here that the controller has real work to do. This is described in detail in Section 6.3.6.

6.3.5 More about Routes

In Section 6.3.1 the class Route was briefly introduced. This class is now revisited, because it is used extensively when calculating how to arrive at the next waypoint.

An instance of Route is an object that represents a journey along a number of Points; see Figure 6.3. It has two properties:

1. No Point is visited more than once in the journey.
2. For each Point p except the last one in the journey, the Point in the journey after p is a *neighbour* of p. A neighbour of a Point p is any Point that is 1 unit in the x or y direction (not both directions) away from p. This is illustrated in Figure 6.4.

These two properties are required for any instance of Route. They therefore need to be described as an invariant on the instance variables in Route. Route has just one instance variable, points, representing this journey:

```
class Route
...
instance variables

points: set of Point;
end Route
```

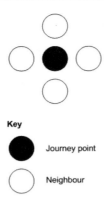

Fig. 6.4: Neighbours of a journey point.

Consider the first requirement. First, it could be rephrased as: For any member p1 of points, there is no other member p2 of points that has the same coordinates. Inverting this line of reasoning, it could again be rephrased as: For any member p1 of points, if p2 is also a member of points and p1 and p2 have the same coordinates, then p1 and p2 must in fact be the same object. This could be written using a forall expression as follows:

```
forall p1, p2 in set points &
   p1.GetCoord() = p2.GetCoord() => p1 = p2
```

For the second property, it is necessary to add an operation to Point to create the neighbours of a Point. From Figure 6.4 it is clear that if the point in question has coordinates mk_(x,y), then there are two potential neighbours that are a vertical step away and two potential neighbours that are a horizontal step away. Because co-ordinates are of type nat * nat, if, for example, x = 0, then it has no neighbour to the left. For all neighbours, the index will be one greater than the index for the current Point. Recalling that x, y and index are instance variables of Point, then neighbours a vertical step away are given by the set:

```
{new Point(x, y1, index + 1) | y1 in set {y - 1, y + 1}
 & y1 >= 0}
```

Similarly the following expression gives the neighbours a horizontal step away:

```
{new Point(x1, y, index + 1) | x1 in set {x - 1, x + 1}
 & x1 >= 0}
```

Having obtained these two sets, the total set of neighbours will be obtained by putting these two sets together to make one set containing the members in each of the two sets. The set operator `union` is provided for precisely this purpose. For instance,

```
{1, 1, 2, 3} union {1, 2, 5, 6, 9}
```

yields $\{1, 2, 3, 5, 6, 9\}$.

The operation `Neighbour` is therefore:

```
class Point
...
operations

public Neighbour: () ==> set of Point
Neighbour () ==
  return {new Point(x, y1, index + 1)
         | y1 in set {y-1,y+1}
         & y1 >= 0} union
         {new Point(x1, y, index + 1)
         | x1 in set {x-1,x+1}
         & x1 >= 0};
end Point
```

Exercise 6.7 Calculate the following set union expressions:

1. {<Apple>, <Orange>} union {<Banana>, <Apple>, <Grape>}
2. {1,...,10} union {}
3. {{1, 2, 3}, {4, 5}} union {{7, -1}, {}}

□

Exercise 6.8 The distributed union operator `dunion` takes a set of sets and returns the union of all of these sets. For example,

```
dunion {{1, 2, 3}, {2, 3, 5}, {}, {7,...,11}}
```

yields $\{1, 2, 3, 5, 7, 8, 9, 10, 11\}$.

Calculate the distributed union in the following cases:

1. dunion {{<Apple>, <Orange>},
   {<Grape>, <Cherry>}, {<Pineapple>}}
2. dunion {}
3. dunion {{}}

□

Exercise 6.9⋆ Without using the dunion operator, write a function that takes a set of sets of natural numbers and returns their distributed union. *Hint:* Use recursion.
□

Returning to the second property of Route, for a member p of points, the property is that the coordinate of the Point after p is a member of the set of coordinates of neighbours of p. This latter set is obtained by the following set comprehension expression:

```
{n.GetCoord() | n in set p.Neighbour()}
```

Assuming an operation called GetNext, which gives the next point in the journey, the property can be formulated as:

```
forall p in set points &
  GetNext(p).GetCoord() in set
  {n.GetCoord() | n in set p.Neighbour()}
```

Note that GetNext allows abstraction away from a particular coordinate system. GetNext has a straightforward definition and is therefore not described further.

There is only one problem with this formulation: The last point in the journey cannot have a next point in the journey! Thus, this condition should apply to all but the last point in the journey. The last point in the journey can be identified as that with index corresponding to the number of points in the journey. Adding this gives:

```
forall p in set points &
  p.GetIndex() <> card points
  => GetNext(p).GetCoord() in set
     {n.GetCoord() | n in set p.Neighbour()}
```

Exercise 6.10 Define the operation GetNext described earlier. □

6.3.6 Avoidance Routes

Returning to the operation PlotCourse introduced in Section 6.3.4, the problem is to plot a course to the next waypoint for a given set of obstacles. The problem is broken down into two subproblems: First, generate a set of possible routes; second, choose one of these routes that is physically acceptable in the sense of requirement

R4 and minimal with respect to distance. If no route is possible, then `nil` is returned by the operation:

```
class Controller
...
operations

PlotCourse: set of (nat * nat) ==> [Route]
PlotCourse(obstacles) ==
 let nextWaypoint = Route'GetPointAtIndex(journeyPlan, 1),
     posRoutes = Route'AvoidanceRoutes(obstacles,
                                       ins.GetPosition(),
                                       nextWaypoint)
 in
   if posRoutes = {}
   then return nil
   else ShortestFeasibleRoute(posRoutes);
end Controller
```

The operation `ShortestFeasibleRoute` is not described further here; instead `AvoidanceRoutes` is described in detail because it represents the critical part of the algorithm.

The function `AvoidanceRoutes` takes three parameters: a set of `obstacles`, which are to be avoided in the route; the `currentPosition` of the robot and the `nextWaypoint` to be visited. Its job is to return the finite set of `Routes` starting at `currentPosition` and finishing at `nextWaypoint`, without colliding with any member of `obstacles`.

How can this be modelled? Recall that the primary objective of the model is to describe *what* is to be done, not *how* it is to be done. In this instance, it would be easy to be distracted by the details of how to model an algorithm for `AvoidanceRoutes`; choosing such a path would lead to a large and complex model. On the other hand, expressing *what* is to be done, can be formulated quite concisely. `AvoidanceRoutes` is therefore modelled by an implicit function definition. The function has no precondition; in the postcondition the desired relationship between the result (called `routes`) and the parameters is modelled. Rephrasing the description in the preceding paragraph, it is required that for each member `r` of `routes`, three properties hold:

1. The starting point in `r` must have the same coordinates as `currentPosition`.
2. The last point in `r` must have the same coordinates as `nextWaypoint`.
3. None of the coordinates of `r` should also be in the set `obstacles`.

The first two properties are straightforward to formulate in VDM++. Assuming operations `GetFirst` and `GetLast` in `Route` this gives

```
r.GetFirst().GetCoord()  = currentPosition.GetCoord() and
r.GetLast().GetCoord()   = nextWaypoint.GetCoord()
```

For the last property, the coordinates of r can be obtained by r.GetCoords():

```
class Route
...
operations

GetCoords: () ==> set of (nat * nat)
GetCoords() ==
  return {p.GetCoord() | p in set points};
end Route
```

The property can then be stated as "r.GetCoords() and obstacles have no members in common" or even "the set of members common to both r.GetCoords() and obstacles is empty." VDM++ provides the operator inter (short for "intersection") to obtain the set of common members. That is, s1 inter s2 yields the set of values that are members of both s1 and s2. For example,

```
{1, 1, 2, 3} inter {1, 2, 5, 6, 9}
```

yields the set {1, 2}. If the two sets have no common members, the intersection is empty. For example

```
{1, 1, 2, 3} inter {4, 6, 9, 11}
```

yields the empty set.
    Property 3 can then be expressed as:

```
r.GetCoords() inter obstacles = {}
```

Putting the three properties together gives the following function:

```
class Route

...

functions

static
public AvoidanceRoutes(
                obstacles: set of (nat * nat),
                currentPosition: Point,
                nextWaypoint: Point) routes: set of Route
post forall r in set routes &
        r.GetFirst().GetCoord() =
        currentPosition.GetCoord() and
        r.GetLast().GetCoord() =
        nextWaypoint.GetCoord() and
        r.GetCoords() inter obstacles = {};
end Route
```

It is possible for this function to be defined as static because it does not use any of the class's nonstatic members. It is convenient to declare it as static because it allows easy access from other classes.

Exercise 6.11 Define the operations GetFirst and GetLast. *Hint:* Use the operation GetPointAtIndex. □

Exercise 6.12 Calculate the following set intersection expressions:

1. {<Apple>, <Orange>} inter
   {<Banana>, <Apple>, <Pineapple>}
2. {1,...,10} inter {}
3. {{1, 2, 3, 4}, {4, 5}} inter {{4, 7, -1}, {4}}

□

Exercise 6.13 The distributed intersection operator dinter takes a non-empty set of sets and returns the intersection of all of these sets. For example

```
dinter {{1, 2, 3}, {2, 3, 5}, {2}, {1,...,11}}
```

yields {2}.

Calculate the distributed intersection in the following cases:

1. dinter {{<Apple>, <Orange>},
   {<Grape>, <Cherry>, <Orange>},
   {<Orange>, <Pineapple>}}

2. dinter {{<Apple>, <Orange>},
   {<Grape>, <Cherry>, <Orange>},
   {<Orange>, <Pineapple>},{}}
3. dinter {{}}

□

Exercise 6.14⋆ Without using the dinter operator, write a function that takes a nonempty set of sets of natural numbers and returns their distributed intersection. *Hint:* use recursion. □

Note that an alternative way of formulating this last property would have been to use the operator not in set. This expresses the opposite of in set. That is,

```
x not in set s
```

yields false if x in set s and true otherwise. For example,

```
1 not in set {1, 1, 2, 3}
```

yields false whereas

```
<Apple> not in set {<Pineapple>, <Orange>}
```

yields true.

Property 3, that none of the coordinates of r should be in the set obstacles, could be written as:

```
forall c in set r.GetCoords() & c not in set obstacles
```

Exercise 6.15⋆ The specification of AvoidanceRoutes is too permissive. It could be implemented by a function that always returned {}, because that would satisfy the postcondition, albeit trivially. Equally, there is no guarantee that all routes are returned. Modify the function so that if it is possible to find a route; then the function must return a nonempty set. *Hint:* Use an exists expression. □

6.3.7 Review

Before continuing with the model it is appropriate to review our progress so far. Having listed the requirements for the robot controller, a representation of the classes

Point and Route, used to represent physical locations, was given. Having modelled these classes, they were then used to model a robot's journeyPlan. Once these classes had been captured, it was possible to begin modelling the actual functionality of the robot controller. The behaviour of the robot was then modelled by decomposing the original requirements and reformulating them in terms of sets.

6.3.8 Extending the Robot Controller

Having described the basic robot controller, a simple extension is now discussed.

The extension is that when selecting a route out of all the possible ones found, preference should be given to those routes that actually include a specific set of coordinates. This might be, for example, that there are nonessential tasks that the robot could perform at these coordinates, but failure to perform these tasks would not compromise the robot's overall objective.

The core of this extension is to be able to model the property that a route visits a specific set of coordinates. The operation VisitsCoordinates in the Route class will model this property. It takes a parameter – the set of coordinates to be met coords – and returns true if all of these coordinates are visited by this route. The requirement to be formulated is that each member of coords is also a member of the set of coordinates of this route (given by GetCoords()). The set operator subset provides precisely this functionality: The expression

```
s1 subset s2
```

yields true exactly when each member of s1 is also a member of s2. For instance,

```
{1, 1, 2} subset {1,...,10}
```

yields true. But

```
{1,...,10} subset {1, 1, 2}
```

yields false.

VisitsCoords would therefore look like:

```
class Route
...
VisitsCoords: set of (nat * nat) ==> bool
VisitsCoords(coords) ==
   return coords subset GetCoords();
end Route
```

Exercise 6.16 Calculate the following subset expressions:

1. $\{1, \ldots, 10\}$ subset $\{1, \ 10\}$
2. $\{1, \ 10\}$ subset $\{1, \ldots, 10\}$
3. $\{$<Apple>, <Orange>, <Cherry>$\}$ subset
   $\{$<Apple>, <Orange>, <Cherry>$\}$

□

Exercise 6.17 Without using the subset operator, define a function that takes two sets of natural numbers and returns true if the first set is a subset of the second set. *Hint:* Use a forall expression. □

Exercise 6.18 The "proper subset" operator psubset takes two sets and returns true if and only if the first set is a subset of the second set *and* the two sets are not equal.

Calculate the following psubset expressions:

1. $\{1, \ldots, 10\}$ psubset $\{1, \ 10\}$
2. $\{1, \ 10\}$ psubset $\{1, \ldots, 10\}$
3. $\{$<Apple>, <Orange>, <Cherry>$\}$ psubset
   $\{$<Apple>, <Orange>, <Cherry>$\}$

□

Exercise 6.19 The set power operator power takes a set and returns the set of all finite subsets of that set. For example,

---
power $\{$<Apple>, <Orange>$\}$

---

yields the set $\{\{\}, \{$<Apple>$\}, \{$<Orange>$\}, \{$<Apple>, <Orange>$\}\}$.
   Calculate the following expressions using power:

1. power $\{1, \ 2, \ 4\}$
2. power $\{$<Banana>$\}$
3. power $\{\}$

□

Exercise 6.20★ Without using the power operator, define a function that takes a set of natural numbers, and returns the set of all finite subsets of this set. *Hint:* Use recursion. □

6.4 Summary

Sets are finite, unordered collections of values in which repetition is not significant. They provide a powerful means of abstraction for modellers, allowing data and algorithms to be modelled without the need to define an arbitrary order. In this way maximum flexibility can be preserved in the model, postponing commitment to a particular order for the data and algorithms, until totally necessary.

The type constructor and operators for sets have been described and demonstrated in this chapter. The operators allow convenient manipulation of set values, allowing the modeller to concentrate on the salient features of the problem in hand, without needing to deal with the low-level details of how the collections are implemented.

The robot controller example presented in this chapter demonstrated the power of using sets. An overall large and complicated navigation system was systematically decomposed into small manageable pieces, which could be modelled using sets. The power of this abstraction approach was underlined by the implicitly defined function `AvoidanceRoutes`, in which sets were used to be able to precisely state the desired properties of the function, without the need to give a complicated algorithm. An alternative approach would have been to use explicit tuples rather than using `Point` objects. The advantage of the object-based approach is that it is more flexible, making it easier to, for example, add new functionality to `Point`.

It can be argued that the use of `Point` objects with indices means that the set of `Point` objects is essentially ordered, because the index values give an ordering. In the next chapter the issue of modelling ordered collections is explored. Recasting this example using an ordered collection is therefore presented as an exercise in that chapter.

7

Modelling Ordered Collections

This chapter aims to introduce the sequence type constructor for modelling ordered collections. On completion of the chapter the reader should understand the defining properties of sequences, be able to use the basic sequence operators, and understand how sequences can be used appropriately in a large model.

7.1 Introduction

Chapter 6 introduced sets as a fundamental data abstraction, representing collections of values in which ordering and repetition are not significant. Many situations, however, demand the use of a data abstraction for collections in which ordering and repetition do matter. VDM++ provides the powerful *sequence* data abstraction for such situations. This chapter begins with an introduction to sequences and then uses an example (a congestion warning system) to demonstrate their use.

7.2 Overview of Sequences

A sequence is an *ordered collection of data* of arbitrary length. Sequence values are represented using using square brackets ('[' and ']') as delimiters, with individual elements separated by commas. The elements are ordered from left to right. For example, the value

```
[1, 2, 3, 3]
```

is a sequence consisting of four values: the natural numbers 1, 2, 3 and 3 again in that order. The type of this value is, therefore, *sequence of natural numbers*, in VDM++ written as follows:

```
seq of nat
```

Sequence types are defined as follows in VDM++:

```
seq of Type
```

This defines the type of all finite sequences of values drawn from the type *Type*. It includes the *empty sequence*, which is represented by the following symbol:

```
[]
```

A special type constructor is used for nonempty sequences. It takes the following form:

```
seq1 of Type
```

For example, the empty sequence [] is of type `seq of` X and not of type `seq1 of` X, for any type X.

The order in which the elements in a sequence appear is relevant, even if the sequence 'contains' the same elements. For example, the following sequence

```
[1, 2, 3, 3]
```

is different from

```
[2, 1, 3, 3]
```

because the elements do not appear in the same order in the two sequences. Similarly, the presence of duplicates is significant, so the following sequence

```
[1, 2, 3]
```

is different from

```
[1, 2, 2, 3]
```

Exercise 7.1 Which of the following pairs of sequences are equal?

1. [1, 2, 3] and [3, 2, 1]
2. [1, 2, 1] and [1, 2, 1]

3. [] and [[]]
4. [] and []

□

As with sets, sequences can be defined by enumeration or by comprehension. The preceding examples all use enumeration, in which the elements are listed in order from left to right. Sequence comprehension makes it possible to define a particular sequence less explicitly. A sequence comprehension has the following form:

```
[create a value using i | i in set numeric set & test i]
```

Such an expression defines a sequence as follows:

1. Numeric values are taken from the set of numeric values *numeric set* in order, smallest first and bound to an identifier such as i.
2. Each binding of i is checked against an (optional) *test* predicate. If the test yields false, that value for i is ignored. If the test yields true, the binding is used in *value creation* defined in front of the vertical bar, and the value resulting from that is added as the next element of the sequence.

This is simpler than it appears. For example, the following sequence comprehension

```
[i*i | i in set {1,..., 10} & i mod 2 = 0]
```

would be evaluated as follows. First the value 1 would be taken from the set {1, . . . , 10}, but because 1 mod 2 is not equal to 0, the result would not be taken into account for the evaluation of i*i. Next, the value 2 would be taken from the set {1, . . . , 10}, and because 2 mod 2 equals 0, 2 would be taken into account for the evaluation of i*i, and hence the value 2*2 = 4 would be the first element of the sequence, and so forth. The final result would be:

```
[4, 16, 36, 64, 100]
```

Exercise 7.2 Calculate the following sequence comprehensions:

1. [i | i in set { 1, 2, 4, 5 } & i < 1]
2. [i | i in set { 1, 2, 4, 5 } & i = 3]
3. [{i, i**2} | i in set { 1, 2, 4 } & i = 4 or i = 1]
4. [[i] | i in set { j | j: nat & j <= 5 }]

□

Exercise 7.3 Write down a sequence comprehension that evaluates to the following:

```
[5,  4,  3,  2,  1]
```

□

VDM++ has 11 operators for use with sequences, summarised in Table 7.1. The most fundamental is sequence application, sometimes called sequence lookup. Sequence application is written in the same way as function application:

```
s(i)
```

A sequence application takes a sequence s and an index i, where i must be in the indices of the sequence, and returns the ith element of s. In VDM++, sequences are indexed from 1 up. A sequence application using an index not in range for the sequence is undefined, yielding a run-time error. For example, the following sequence application

```
[2,  4,  6,  8,  10](3)
```

yields the value 6. Suppose sequence s has the value

```
[<red>,  <green>,  <blue>]
```

Then the following sequence application

```
s(2)
```

yields the value <green>, but the following expression

```
s(6)
```

is undefined.

The length of a sequence is given by the sequence length operator. This operator takes a single sequence, say s, as its argument and returns the current number of elements in that sequence. It is written as follows:

```
len s
```

For example, again considering the sequence s having the value

```
[<red>,  <green>,  <blue>]
```

the following expression

```
len s
```

yields the value 3 and

```
s(len s)
```

yields the value <blue>. The valid indices of a sequence range from 1 up to the sequence length.

Exercise 7.4 Evaluate the following expressions:

1. len [4, 5, 3, 3]
2. len [{4, 5, 3, 3}]
3. len []
4. len [[]]

□

Other important sequence operators are the *head* (hd) and *tail* (tl) operators. The head operator takes a sequence as its argument and returns the first element of that sequence as its result. So the expression

```
hd [4, 16, 32, 64, 100]
```

yields the value 4. The head operator is the same as performing a sequence application on index 1. As a result, the hd operator is partial; if applied to an empty sequence it results in an undefined value. Attempts to do so in, e.g., operation definitions must be avoided, for instance by using preconditions. Note also that the head operator does not yield a sequence but a value of the type of which the sequence is made.

The tail operator is the opposite of the head operator: Instead of returning the value of the first element it returns the sequence consisting of everything but the first element. So the expression

```
tl [4, 16, 32, 64, 100]
```

yields the value

```
[16, 32, 64, 100]
```

Again, this operator is partial; it is undefined if applied to an empty sequence.

Exercise 7.5 Evaluate the following expressions:

1. `hd [true, false, true]`
2. `hd [[true, false], [true]]`
3. `tl [true]`
4. `tl [[true], [false, true]]`

□

The remaining sequence operators are given in Table 7.1 and are introduced through a larger example in the next section of this chapter, which describes the use of sequences in a realistic setting.

Table 7.1: Summary of VDM++ sequence operators

Operator	Name	Type
`hd l`	Head	`seq1 of A → A`
`tl l`	Tail	`seq1 of A → seq of A`
`len l`	Length	`seq of A → nat`
`elems l`	Elements	`seq of A → set of A`
`inds l`	Indexes	`seq of A → set of nat1`
`l1 ^ l2`	Concatenation	`(seq of A) * (seq of A)` `→ seq of A`
`conc ll`	Distributed concatenation	`seq of seq of A → seq of A`
`l ++ m`	Sequence modification	`seq of A * map nat1 to A` `$\xrightarrow{\sim}$ seq of A`
`l(i)`	Sequence application	`seq of A * nat1 $\xrightarrow{\sim}$ A`
`l1 = l2`	Equality	`(seq of A) * (seq of A) → bool`
`l1 <> l2`	Inequality	`(seq of A) * (seq of A) → bool`

7.3 The Congestion Warning System (CWS)

Traffic can be dense on motorways, and as a result congestion can occur. To enhance safety on the road, congestion warning systems can be used to warn drivers of upcoming congestion, suggesting that they reduce speed so that chances of collisions at the rear of the congestion are minimized.

The general structure of a traffic management system is illustrated in Figure 7.1. The main components are as follows:

Sensors are used to gather information on the current behaviour and status of the traffic. Examples of sensors are video cameras and radar but even humans are used in some cases to gather data for the appropriate authorities. Sensors can be defined according to layered models, where data provided by lower-level sensors (e.g., a loop detector) are aggregated by a higher-level sensor (e.g., average speed).

TrafficControls are used to interpret the data coming from sensors and take appropriate action such as warning of congestion, ramp metering or diverting traffic.

Actuators are used to convey signals to drivers based on the action required by Traffic-Controls. Actuators exist in many forms, either close to the roads, where they can be seen by car drivers (variable message signs, traffic lights), or in cars themselves (e.g., intelligent speed adaptation). A special problem here is that the signals conveyed to cars must be consistent and coherent; otherwise road safety is decreased instead of increased. For this purpose an *actuator manager* can be used.

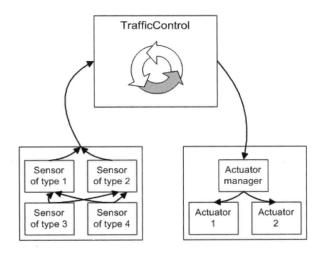

Fig. 7.1: General structure of traffic management systems

The CWS is built according to these principles. It "senses" at a number of points whether congestion has occurred. A variety of sensors could be used for this; loop detectors (explained later) are used in the example. The CWS transforms the sensor data into signals, which are transmitted to drivers using actuators. Again, there are many types of possible actuators, including devices inside the cars themselves, but in the example, simple illuminated road signs are used. The one in Figure 7.2(a) indicates congestion by imposing a speed limit of 50 km/h. The one in Figure 7.2(b) indicates congestion ahead and a "blank" sign implies that the traffic is fluid and the situation is normal. Figure 7.3 presents an overview of the CWS system in which the leftmost actuator is blank, indicating that there is no congestion.

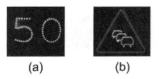

(a)   (b)

Fig. 7.2: Simple traffic signs: (a) speed limitation and (b) congestion announcement

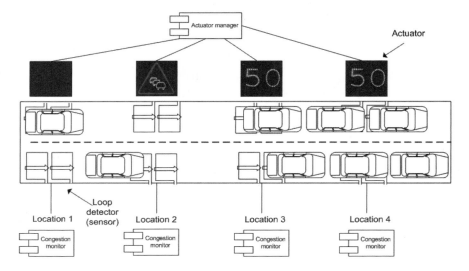

Fig. 7.3: Overview of the CWS system

The CWS system is now being built up step by step, in a top-down fashion. First, the class CWS is introduced. The overall structure of the class diagram in which CWS is at the centre is given in Figure 7.4.

At the heart of the example is the concept of a "road network." The road network in the example is quite abstract: A motorway simply consists of a number of consecutive points where congestion is being monitored. Because the order of congestion monitors is important here, we use the VDM++ *sequence* data type, which models exactly that notion:

```
roadNetwork: seq of CongestionMonitor
```

Note that in the UML class diagram in Figure 7.4 this is shown as an ordered association. Sequence indexing means that it is possible to use sequence application to access a specific element in the sequence. For example, the expression

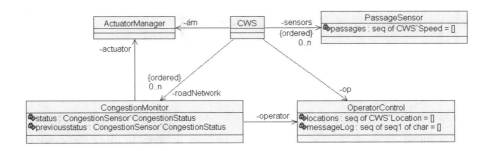

Fig. 7.4: UML class diagram for CWS

```
roadNetwork(i)
```

would yield the ith element of the sequence `roadNetwork`, a reference to an instantiation of class `CongestionMonitor`. Recall that sequence indexing starts at 1. The `inds` operator yields a sequence's set of indices. So the expression

```
inds roadNetwork
```

will return the set of valid indices for the sequence `roadNetwork`. Assuming, for example, that four congestion monitors were put in place in `roadNetwork`, this expression would yield the value $\{1,2,3,4\}$.

Exercise 7.6 The following function takes the sequence s and the number n as inputs. It filters out the elements of s that are smaller than n. Complete its definition using sequence comprehension and the `inds` operator:

```
FilterSmall: seq of int * int -> seq of int
FilterSmall(s, n) ==
```

☐

It is possible for a road network to have no congestion monitors at all, and therefore the sequence defined here is the kind that may be empty (`seq of`). Initially, the sequence is empty:

```
class CWS
...
instance variables

roadNetwork: seq of CongestionMonitor := [];
end CWS
```

In much the same way, a number of sensors are defined. In this case sensors of class PassageSensor are used. To ensure that for each congestion monitor there is a corresponding sensor that it can use, an invariant is given that expresses the condition that the length of the sequence modelling the congestion warning objects is the same as the length of the sequence modelling the available sensors. For this purpose the sequence operator len is used, which takes a sequence as its argument and returns the *length* of that sequence in terms of the number of elements in that sequence:

```
class CWS
...
instance variables

roadNetwork: seq of CongestionMonitor := [];
sensors    : seq of PassageSensor := [];
inv len roadNetwork = len sensors
end CWS
```

Finally, two objects representing an actuator manager and an operator control are declared. The actuator manager is responsible for distributing signals to the various actuators in the system. The operator control provides several facilities for monitoring the system from, e.g., a traffic centre:

```
class CWS
...
instance variables

am: ActuatorManager := new ActuatorManager();
op: OperatorControl := new OperatorControl();
end CWS
```

Several types need to be introduced. The most interesting, from the perspective of the example, is the type Signal, modelling the signals that can be conveyed by the actuators in the example. Because the signals are discrete and the relative order between the values is of no importance, the use of a union of quote types is best used here:

```
class CongestionMonitor
...
types

public Signal = <NoWarning>|<PreAnnouncement>|
                <CongestionWarning>;
end CongestionMonitor
```

Because the system initially has no congestion warning objects, a facility must be provided to populate the system. This is done by defining an operation for adding a new congestion monitor called AddCongestionMonitor. This involves introducing a sensor to go with it. In this case it is assumed that congestion detection is done on the basis of calculating the average speed of passing vehicles; hence a sensor of class PassageSensor needs to be instantiated first.

Both the congestion monitor and the sensor are added at the appropriate place in the network. This means that the instance variables roadNetwork and sensors need to be updated. Two sequence operators are needed in this situation: the *subsequence expression* and the *sequence concatenation operator*.

The subsequence operator creates a subsequence of a sequence s between a lower bound lb and an upper bound ub. The subsequence operator is written as

```
s(lb,...,ub)
```

For example, suppose that a sequence s has the value

```
[1, 3, 5, 7, 9, 11]
```

Then the expression

```
s(3,...,5)
```

yields the value

```
[5, 7, 9]
```

There are no constraints on the lower and upper bounds for this operator. If the upper bound is smaller than the lower bound, then the empty sequence is returned. If the lower bound is smaller than 1, then a lower bound value of 1 is assumed. If the upper bound is larger than the length of the sequence, then an upper bound equal to the length of the sequence is assumed.

Exercise 7.7 Evaluate the following expressions:

1. [4, 5, 3, 3, 9, 3, 2, 3](2, ..., 5)
2. [4, 5, 3, 3, 9, 3, 2, 3](5, ..., 12)
3. [4, 5, 3, 3, 9, 3, 2, 3](5, ..., 5)
4. [4, 5, 3, 3, 9, 3, 2, 3](5, ..., 2)

□

Sequence concatenation is one of the more important sequence operators in the sense that it is regularly used in models involving sequences. The operator is written using the circumflex symbol (""). It takes two sequences s1 and s2 as its arguments and returns the concatenation of these sequences as its result. For example, the expression

```
[1, 2, 3] ^ [3, 2, 1]
```

yields

```
[1, 2, 3, 3, 2, 1]
```

and

```
[] ^ [1, 2, 3] ^ [] ^ [1, 2, 3] ^ [3, 2, 1]
```

yields

```
[1, 2, 3, 1, 2, 3, 3, 2, 1]
```

Exercise 7.8 Evaluate the following expressions:

1. [] ^ [] ^ []
2. [true, false] ^ [hd [false, true]]
3. [true, false] ^ [tl [false, true]]
4. [] ^ [[] ^ []]

□

The two operators are used to construct new values for both roadNetwork and sensors. Because sequence concatenation takes sequences as arguments, explicit sequence enumeration is used to construct a sequence from the single entities cm and sensor, written as [cm] and [sensor], respectively.

It is important to remember here that an *invariant* was defined, constraining the relationship between roadNetwork and sensors. This invariant could be broken during the execution of AddCongestionMonitor if changes are made to either instance variable. Because operations in general can rely on the invariant holding and operations can be invoked in each statement, the invariant must be preserved by each statement executed.

To resolve this problem, VDM++ has a *multiple assignment statement*. This statement is written as:

```
atomic (assignment statement 1;
        assignment statement 2;
        ...
        assignment statement n
       )
```

The effect of this statement is that all statements *assignment statement 1* through *assignment statement n* are executed sequentially, but no evaluation of the invariant occurs during execution. Hence, if the invariant holds before the execution of the multiple assignment statement, the component assignments should together ensure that the invariant holds after execution of the statement. The definition for the operation AddCongestionMonitor then becomes as follows:

```
class CWS
...
operations

public AddCongestionMonitor: Location ==> ()
AddCongestionMonitor(loc) ==
(def sensor = new PassageSensor(loc);
     cm = new CongestionMonitor(loc, sensor, am, op)
 in
   let numberOfWarners = len roadNetwork
   in
     atomic(roadNetwork := roadNetwork(1,...,loc) ^
                           [cm] ^
                           roadNetwork(loc+1,...,
                                       numberOfWarners);
            sensors := sensors(1,...,loc) ^ [sensor] ^
                       sensors(loc+1,...,
                               numberOfWarners)
           );
```

```
  am.AddActuator(loc)
)
end CWS
```

where the Location is defined to be positive natural numbers. Note that the final call of the AddActuator adds an actuator to convey signals where the congestion monitor has been added.

7.3.1 Sensors

In the sensor part of the CWS system, the general principles for constructing traffic management systems are followed in the sense that a "layered model" is constructed using inheritance. This is shown in the UML class diagram in Figure 7.5.

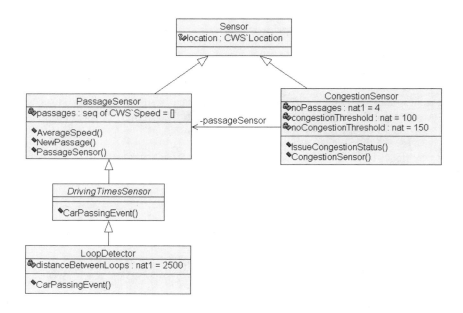

Fig. 7.5: UML class diagram for sensors

At the top of the inheritance tree the general notion of a "sensor" is captured. The only relevant information being modelled here is the notion that every sensor has a location. All other relevant information, such as what the sensor is actually sensing and how it goes about doing this, is left to the subclasses. In the example, at the lowest level a loop detector is used. A loop detector consists of two induction loops placed under the road, at a given distance from one another, as illustrated in Figure 7.6.

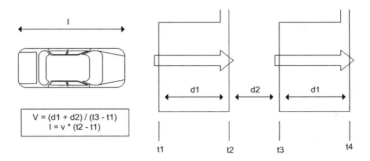

Fig. 7.6: Principle of a loop detector

A loop detector can detect a single vehicle passage and calculate the speed of the vehicle. A passage sensor is then capable of calculating an average speed from a predefined number of vehicle passages. In the example, this information is used by a congestion sensor to determine whether there is congestion at the location of the sensor. Hence, a congestion sensor is modelled as a subclass of Sensor (after all it is a sensor itself) and at the same time it uses a PassageSensor, modelled using a UML association to get the information needed to determine whether congestion has occurred.

In VDM++ terms, all this can be expressed as follows. First, consider the class Sensor. The instance variable location is defined as protected so that it can be modified by instances of its subclasses:

```
class Sensor

instance variables

protected location: CWS'Location

end Sensor
```

Class PassageSensor needs to capture information about the speed of a number of vehicle passages. In this case, again, a sequence is most appropriate. A set would not be suitable because, although the order of vehicle passages does not matter for calculating the average speed, the fact that there may be duplicates in the sequence is relevant:

```
class PassageSensor is subclass of Sensor
...
instance variables

passages: seq of CWS'Speed := []
end PassageSensor
```

Notice that the instance variable *location* is not initialised directly in the super-class of `PassageSensor`: class `Sensor`. The reason is that when an object from this class is instantiated, it has to be placed at a certain location. Hence, an explicit constructor is needed that provides the newly instantiated object with that information:

```
class PassageSensor is subclass of Sensor
...
operations

public PassageSensor: CWS'Location ==> PassageSensor
PassageSensor(loc) ==
  location := loc;
end PassageSensor
```

The first operation needed for this class is one that can add a vehicle passage to its instance variable `passages`. The method is called `NewPassage` and again an example of sequence concatenation can be seen. New vehicle passages are added to the front of `passages` because only a predefined number of vehicle passages are taken into account when calculating the average speed, needed for congestion detection. Having the most recent passages at the front of the sequence is more convenient when capturing them in the process of calculating the average:

```
class PassageSensor is subclass of Sensor
...
public NewPassage: CWS'Speed ==> ()
NewPassage(speed) ==
  passages := [speed] ^ passages;
end PassageSensor
```

The operation `AverageSpeed` needs to calculate the average speed based on a predefined number of vehicle passages. This predefined number can, for example, be called `numberOfPassages`, and it is passed as an argument to the operation. The first thing to do is to capture the relevant vehicle passages into a new sequence.

The easiest way to do this is to use the subsequence operator to capture the first numberOfPassages elements of passages:

```
passages(1,..., numberOfPassages)
```

Notice that the earlier design decision to put the most recent vehicle passages in front of passages makes this relatively easy.

To calculate average speed, an iteration is needed over the elements of the constructed subsequence, summing the elements in the process and dividing by number-OfPassages at the end. For this purpose, a *sequence for loop statement* is available in VDM++. It has the general form:

```
for pattern in sequence do
    statement
```

Iteration starts at the first element of *sequence*, traversing the sequence and performing *statement*, typically involving *pattern* each time.

Accumulation of the speeds of individual vehicle passages can be written as

```
for speed in passages(1,..., numberOfPassages) do
    accSpeed := accSpeed + speed;
```

Before operation AverageSpeed can be written down, it is necessary to ensure that enough vehicle passages have been accumulated so that an average speed can be calculated. In other words, the length of the sequence that holds the vehicle passages needs to be greater than or equal to numberOfPassages. This is formulated in a precondition:

```
pre len passages >= numberOfPassages
```

The full operation then becomes:

```
class PassageSensor is subclass of Sensor
...
public AverageSpeed: nat1 ==> CWS'Speed
AverageSpeed(numberOfPassages) ==
( dcl accSpeed: CWS'Speed := 0;
  let passInAccount = passages(1,..., numberOfPassages)
  in
  ( for speed in passInAccount do
```

```
      accSpeed := accSpeed + speed;
   return (accSpeed/numberOfPassages)
 )
)
pre len passages >= numberOfPassages
end PassageSensor
```

Exercise 7.9 Redefine the operation `AverageSpeed` using a *set for loop* instead of a *sequence loop*. □

Exercise 7.10⋆ Redefine the operation `AverageSpeed` using recursion. *Hint:* Introduce a function. □

The next two more basic levels of sensor are the `DrivingTimesSensor` class and the `LoopDetector` class. Because the speed of vehicle passages (driving times) can be determined in multiple ways (e.g., by using a loop detector and a radar system), the class is defined as an abstract superclass with an operation `CarPassingEvent` defined as

```
class DrivingTimesSensor is subclass of PassageSensor

operations

public CarPassingEvent: nat1 ==> ()
CarPassingEvent(-) ==
   is subclass responsibility

end DrivingTimesSensor
```

The operation models the passing of a car. Because the name of the input parameter (the driving time) is of no importance here, the most general form of a pattern ("-", also referred to as "don't care pattern") is used here as a formal parameter to the operation (see Section 5.5). Then, depending on the specific technology being applied, a subclass can be defined in which a definition for this operation is given:

```
class LoopDetector is subclass of DrivingTimesSensor

values

distanceBetweenLoops: nat1 = 2500

operations
```

```
public CarPassingEvent: nat1 ==> ()
CarPassingEvent(drivingTime) ==
  NewPassage(distanceBetweenLoops * drivingTime)

end LoopDetector
```

The final sensor that needs to be defined is the congestion sensor itself. Based on the average speed being calculated by a more general sensor, it will determine whether congestion has occurred when the average speed comes below a given value (resulting in a status <Congestion>), whether congestion has vanished when the average speed comes above a given value (resulting in a status <NoCongestion>) or whether it cannot be determined if congestion exists (resulting in the value <Doubt>). The algorithm is preferably independent of these threshold values, and these are therefore defined as constant values in the class definition:

```
class CongestionSensor is subclass of Sensor
...
values

congestionThreshold   : nat = 100;
noCongestionThreshold: nat = 150

end CongestionSensor
```

Again, it is desirable to define this class so that it is independent of the specific technology being used to collect average speeds. Instead of defining an association between this class and LoopDetector, an association is defined with the class PassageSensor. It is then up to the overall CWS system to supply a congestion sensor with an instance of a specific subclass of passage sensor, capable of performing the desired tasks. This also means that in class CongestionSensor an explicit constructor needs to be defined to initialize the instance variable passageSensor with an appropriate value:

```
class CongestionSensor is subclass of Sensor
...
instance variables

passageSensor: PassageSensor

types

public CongestionStatus = <Congestion>|<NoCongestion>|
                          <Doubt>
```

```
operations

public CongestionSensor: PassageSensor ==>
                        CongestionSensor
CongestionSensor(sensor) ==
  passageSensor := sensor;

public IssueCongestionStatus: () ==> CongestionStatus
IssueCongestionStatus() ==
  def averageSpeed = passageSensor.
                        AverageSpeed(noPassages)
  in
    if averageSpeed < congestionThreshold
    then return <Congestion>
    elseif averageSpeed > noCongestionThreshold
    then return <NoCongestion>
    else return <Doubt>
end CongestionSensor
```

7.3.2 Traffic Control

At the core of each traffic management system resides a control component, which takes the inputs coming from one or more sensors, performs calculations on these inputs determining whether some action should be taken and then instructs one or more actuators (via an actuator manager) to convey appropriate signals to the road side or to vehicles directly. Many such "traffic control" components can exist. The similarity between them is that they all operate at a given location. This suggests the introduction of a class `TrafficControl`, with an instance variable modelling its location. The class `CongestionMonitor` can then be defined as a subclass of the class `TrafficControl`. Associations are used to connect this class to a congestion sensor to retrieve input data, an actuator manager to which signals are conveyed and an operator control to convey messages to the traffic centre. Furthermore, two instance variables are used to capture the current and previous states of congestion detected by the monitor. The relation between the two classes is given in Figure 7.7.

An explicit constructor is used here to give appropriate initial values to the instance variables.

7.3.3 Actuators

The actuators component of the CWS system has two main parts:

- an actuator manager, which is responsible for controlling a number of actuators and making sure that the signals they convey are coherent and consistent. This

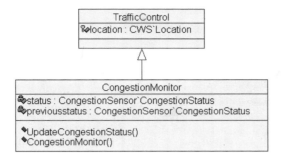

Fig. 7.7: UML class diagram for traffic controls

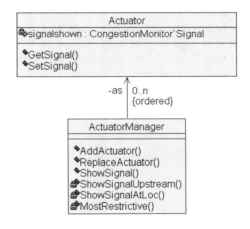

Fig. 7.8: UML class diagram for actuators

means, for instance, that when a congestion warning is shown at location *i*, then a preannouncement must be shown at location *i-1*.

• actuators, which are in a sense "dumb" devices, simply taking orders from an actuator manager and showing the required signals

The relation between an actuator manager and actuators is modelled in UML class diagrams as an ordered association (see Figure 7.8).

At the VDM++ level this means that, in class `ActuatorManager`, an instance variable is defined, modelling a sequence of actuators, initially being empty:

```
class ActuatorManager
...
instance variables

as: seq of Actuator := [];
end ActuatorManager
```

Then a number of operations can be defined, the most important of which shows a given signal at a given location. This operation is called ShowSignal.

In the process of determining which signal needs to be shown where, it is useful to capture the actuators located "downstream" (in the driving direction) and "upstream" (against the driving direction), respectively. More than just the given location may be relevant here. If a congestion warning signal must be shown at location loc, then potentially an announcement sign needs to be displayed at location loc - 1. This means that, from the sequence of available actuators, the appropriate elements must be selected. This can be done using sequence application. In this situation sequence application is used to "identify" the actuator at the desired location, the actuator upstream and the actuator downstream, by writing:

```
class ActuationManger
...
public ShowSignal: CWS'Location *
                    CongestionMonitor'Signal ==> ()
ShowSignal(location, signal) ==
(let downstream = as(location + 1),
     actuator   = as(location),
     upstream   = as(location - 1)
 in
   -- Set the right signal at the location itself
   (ShowSignalAtLoc(signal,downstream,actuator);
   -- Set the right signal upstream

  ShowSignalUpstream(signal,upstream)
 )
)
pre location in set {2,..., len as -1} and
    (signal = <NoWarning> or
     signal = <CongestionWarning>);
end ActuationManger
```

Notice how a signal for the given location is determined, based on the current signal being shown downstream, and that a signal upstream is potentially set, based on the signal shown at the current location. Notice also that, as a result of the need to access actuators potentially out of the range of inds as, a precondition was formulated.

Exercise 7.11 It is slightly unsatisfactory that the precondition prevents signals being set at the borders of the range of ActuatorManager under all circumstances. Reformulate the body of ShowSignal in such a way that the following precondition is valid instead of the current precondition:

```
pre location in set inds as and
    (signal = <NoWarning> or
     signal = <CongestionWarning>);
```

□

Consider the definition of ShowSignalAtLoc:

```
class ActuationManger
...
ShowSignalAtLoc: CongestionMonitor'Signal * Actuator *
                 Actuator ==> ()
ShowSignalAtLoc(signal,downstream,actuator) ==
  if signal = <NoWarning>
  then def downstreamsignal = downstream.GetSignal()
       in
          if downstreamsignal = <CongestionWarning>
          then actuator.SetSignal(<PreAnnouncement>)
          else actuator.SetSignal(<NoWarning>)
  else def currentsignal = actuator.GetSignal()
       in
          let safest = MostRestrictive(currentsignal,
                                       signal)
          in
             actuator.SetSignal(safest);
end ActuationManger
```

A function MostRestrictive is used to make sure that only the "safest" signal is shown. The safest signal in the current set of signals is <CongestionWarning>, because it limits the vehicles more than <PreAnnouncement>. Thus, the signal <CongestionWarning> imposes a speed restriction, whereas this is not the case for the <PreAnnouncement> signal. <PreAnnouncement>, in its turn, is more restrictive than <NoWarning>:

```
class ActuatorManager
...
functions

MostRestrictive: CongestionMonitor'Signal *
                 CongestionMonitor'Signal ->
                 CongestionMonitor'Signal
MostRestrictive(s1, s2) ==
  if s1 = <CongestionWarning> or s2 = <CongestionWarning>
  then <CongestionWarning>
  elseif s1 = <PreAnnouncement> or s2 = <PreAnnouncement>
  then <PreAnnouncement>
  else <NoWarning>
end ActuatorManager
```

Finally for this class, two operations are used to model the addition and replacement of actuator managers. For the addition of a new actuator, as seen before, a combination of the VDM++ subsequence and sequence concatenation is appropriate:

```
class ActuatorManager
...
operations

public AddActuator: CWS'Location ==> ()
AddActuator(loc) ==
  def act = new Actuator()
  in
     as := as(1,...,loc) ^ [act] ^ as(loc+1,..., len as)
pre loc in set inds as;
end ActuatorManager
```

For the replacement of an element in a sequence, VDM++ provides the sequence modifier operator. A sequence modifier takes a sequence s as its argument, an index i to specify where in s an element should be replaced and an expression e denoting the new value of s at index i. This is written as:

```
s ++ {i |-> e}
```

For example, if s is the sequence

```
[<red>, <green>, <blue>]
```

then the expression

```
s ++ {2 |-> <blue>}
```

will yield

```
[<red>, <blue>, <blue>]
```

Note that the index i must be a member of the set inds s.

Exercise 7.12 Calculate the following sequence modifier expressions:

1. [true, false, true] ++ {2 |-> true, 3 |-> 4}
2. let s = [4, 4, 2, 1] in s ++ {4 |-> s(3)}

□

The appropriate expression for the replacement of an actuator in sequence as by another is:

```
class ActuatorManager
...
operations

public ReplaceActuator: CWS`Location ==> ()
ReplaceActuator(loc) ==
 def act = new Actuator()
 in
    as := as ++ {loc |-> act}
pre loc in set inds as;
end ActuatorManager
```

The class Actuator itself is quite simple: It contains an instance variable representing the signal currently shown and two operations for setting a signal and retrieving the currently shown signal, respectively.

7.3.4 Class OperatorControl

Class OperatorControl is used to model the facilities provided to a traffic operator in a traffic centre. Basically, the information an operator gets is based on log messages being sent by congestion monitors at given locations. Each message can be modelled as a nonempty sequence of characters (called a string):

```
seq1 of char
```

whereas the overall log of messages can be modelled as a sequence of messages, i.e.,

```
messageLog: seq of (seq1 of char)
```

It makes no sense for messages to be empty, so the nonempty sequence type constructor is used here. The log itself is initially empty, hence the use of seq of. To
keep track of the locations from which messages originated, the locations are stored
in a separate sequence:

```
locations: seq of CWS'Location
```

To make sure that the two instance variables are internally consistent, an invariant
is used. The instance variable section of the class is written as:

```
class OperatorControl
. . .
instance variables

messageLog: seq of seq1 of char := [];
locations : seq of CWS'Location := [];
inv len messageLog = len locations
end OperatorControl
```

Several other operations, including ResetLog and WriteLog, use sequence
operators already. Notice, though, that WriteLog uses an auxiliary function (left as
an exercise for the reader) to ensure that only elements of the correct type are added
to messageLog:

```
class OperatorControl
. . .
operations

public ResetLog: () ==> ()
ResetLog() ==
  messageLog := [];

public WriteLog: seq1 of char * CWS'Location ==> ()
WriteLog(message, location) ==
(messageLog := messageLog ^
                [message ^ ConvertNum2String(location)];
 locations := locations ^ [location]
```

```
);
end OperatorControl
```

Exercise 7.13⋆ ConvertNumString, here, is a function that takes a numeric value and returns a sequence of (numeric) characters representing that value. Formulate that function. □

Exercise 7.14⋆ Reformulate the invariant for the instance variables in the class OperatorControl so that it includes the notion that message(i) is derived from locations(i). □

Exercise 7.15 Rewrite the operation WriteLog in an implicit manner (i.e., using a postcondition) without using the sequence operators hd and tl. □

An operation CongestionSpots is required, which, when executed, returns a set with all locations at which congestion has occurred at some time. This operation can be used to analyse "weak spots" in the road where congestion occurs, so that measures can be taken to avoid this in the future. This information is available in the instance variable locations, but the sequence locations need to be transformed into a set for this purpose. VDM++ has a sequence operator elems that is suited for this purpose: It takes a sequence s as its argument and returns a set with the same elements. Of course, because a set cannot contain any duplicate elements, any such duplicates available in the sequence will "disappear." This is written as

```
elems s
```

Consider a sequence s

```
[7, 9, 3]
```

Then elems s will yield

```
{3, 7, 9}
```

And consider a sequence s

```
[1, 2, 2, 2, 2, 3, 4]
```

Then elems s will yield

```
{1, 2, 3, 4}
```

Exercise 7.16 Evaluate the following expressions:

1. elems [<Red>]
2. elems [2, 5, 4, 5]
3. elems [tl [2]]
4. elems [tl [2, 5, 4, 5]]

□

In this situation the operation CongestionSpots can simply be written as:

```
class OperatorControl
...
operations

public CongestionSpots: () ==> set of CWS'Location
CongestionSpots() ==
  return elems locations;
end OperatorControl
```

The final operation for this class is ConvertLog2Field which takes a message log and converts it to a "flat" file, so that message can be exported outside the CWS system. Files can simply be modelled as

```
seq of char
```

This means that the message log, which was modelled as

```
seq of (seq1 of char)
```

must be flattened into a "single" sequence.

The VDM++ distributed concatenation operator can be used for this purpose. This operator takes a sequence of sequences ss as its argument and returns a single sequence. It is written as

```
conc ss
```

Suppose ss has the value

```
[[<red>, <green>], [<green>], [<blue>, <orange>]]
```

Then conc ss will yield

```
[<red>, <green>, <green>, <blue>, <orange>]
```

ConvertLog2File can be written using the conc operator:

```
class OperatorControl
...
operations

ConvertLog2File: () ==> seq of char
ConvertLog2File() ==
    return conc messageLog
end OperatorControl
```

Exercise 7.17 Redefine the robot controller as presented in Chapter 6 using sequences instead of sets. □

7.4 Summary

This chapter has presented the sequence type constructor as a means of modelling ordered collections of values in which repetition is significant. While providing an appropriate level of abstraction, sequences at the same time provide mechanisms such as duplication of values and ordering, which can in many situations effectively be used to specify an appropriate model. The various operators for sequences have been described and demonstrated. These operators allow convenient manipulation of sequence values so that the modeller can concentrate on the salient features of the problem at hand.

8

Modelling Relationships

This chapter aims to introduce mappings as a means of modelling relationships between values. The constructs are illustrated using the same example as in Chapter 7, but more advanced modifications to the example are made, requiring the use of the powerful mapping data type. On completion of this chapter the reader should understand the defining properties of mappings, be able to use the basic sequence operators and understand how mappings can be used appropriately.

8.1 Introduction

Ordered and unordered collections are valuable modelling tools. However, many situations demand the use of more sophisticated structures to model relationships between collections, for example, associating identifiers with data values, defining directories or look-up tables. For such situations, VDM++ provides a powerful data abstraction called a *mapping*. In this chapter, the basic mapping constructor and operators are described first, and then a case study introduces the remaining operators. The case study is based on the congestion warning system from Chapter 7. It shows how mappings can be used to model aspects of a realistic CWS that could not be described so readily using only sets and sequences.

8.2 Overview of Mappings

A mapping is a data structure describing a relationship between the values of one set, called the *domain*, and the values of another set, called the *range*. A mapping can be thought of as a table or directory, in which one can look up a domain value and read across to the corresponding range value. This is a good analogy, apart from the fact that table entries are usually ordered in some way, whereas mappings are unordered. We will say that each domain element *maps to* the corresponding range element. There can be no ambiguity about what range element each domain element maps to, so each domain element can occur only once in a mapping.

Mappings are essentially sets of pairs of domain and range elements. Each pair is called a *maplet*) and a mapping is represented using curly brackets ("{" and "}") as delimiters, with individual maplets separated by commas. Each maplet consists of a domain value and a range value separated by the map symbol ("| ->"). For example, the value

```
{"Paul"  |-> 7, "Peter"  |-> 5, "Marcel"  |-> 6,
 "John"  |-> 8, "Nico"  |-> 8}
```

is a mapping consisting of five maplets: {"John" | -> 8}, {"Nico" | -> 8}, {"Marcel" | -> 6}, {"Paul" | -> 7} and {"Peter" | -> 5}. The type of this value is, therefore, *map from strings to natural numbers*, expressed as follows in VDM++:

```
map (seq of char) to nat
```

As with sets, the order in which maplets appear in a mapping is not significant.

The mapping type constructor in VDM++ is as follows:

```
map type1 to type2
```

This denotes the type of all finite mappings between elements of *type1* and elements of *type2*; *type1* and *type2* are called the domain and range types of the mapping, respectively.

Mappings are unambiguous in the sense that each domain value points to just one range value. However, several domain values can point to the same range value if necessary. There are situations in which the domain elements should uniquely identify range elements, so that the mapping is *one to one* and not *many to one*. Such mappings are said to be *injective*. There is a special type constructor for injective mappings:

```
inmap type1 to type2
```

For example, the expression

```
{"John"  |-> 8, "Nico"  |-> 8}
```

is a valid mapping but *not* a valid injective mapping, and

```
{"Peter"  |-> 7, "Peter"  |-> 5}
```

is not a valid mapping because there are two different maplets with the same domain value. The expression

```
{"John"  |-> 9,  "Nico"  |-> 8,  "Peter"  |-> 5}
```

represents a valid injective mapping. The important practical characteristic of injective mappings is that they are "two way"; an injective mapping can be inverted to yield a useful valid mapping. The operator for doing this is discussed later.

The *empty mapping*, which has no maplets, is represented as: {|->}.

Exercise 8.1 Consider each of the following pairs of mappings. Which are equal?

1. {1 |-> 5} and {5 |-> 1}
2. {1 |-> 5, 5 |-> 1} and {5 |-> 1, 1 |-> 5}

□

Mappings can be defined by enumeration or by comprehension. Mapping comprehension is analogous to set comprehension, except that the expression generates sets of maplets instead of sets of single values. It takes the following form:

```
{maplet | bindings & predicate}
```

Such an expression builds up a mapping as follows:

1. Values are taken from *bindings*.
2. These values are evaluated against an (optional) *predicate*.
3. If the predicate yields true, then the value is used to form *maplet*, and that maplet is added to the mapping.

For example, the *mapping comprehension*

```
{i  |-> i*i  |  i: nat1 & i <= 4}
```

would be built up as follows. A value from nat1 is selected and bound to i. Suppose the value 1 is selected. Because the predicate i <= 4 holds for this value of i, the maplet {1 |-> 1} would be added to the mapping. Next the value 2 would be taken for i; the predicate would be evaluated and because it also holds for this case, the resulting maplet {2 |-> 4} would be added to the mapping, and so forth.

The final result would be:

```
{1  |-> 1, 2  |-> 4, 3  |-> 9, 4  |-> 16}
```

Mappings, like sets and sequences, have to be of finite size, and care must be taken with comprehensions to ensure this. Finiteness can be guaranteed if the bindings are set bindings, e.g., `i in set {1, ...5}`, because sets are finite. Alternatively, check that the test predicate is only true for a finite number of values satisfying the binding, as in the preceding example. In this case the set bind `i in set {1, ..., 4}` could have been used as an alternative. See Section 5.6 for more information about bindings.

Exercise 8.2 Calculate the following mapping comprehensions:

1. `{i**2 |-> i/2 | i in set {1, 2, 4, 5}}`
2. `{i**2 |-> i/2 | i in set {1, 2, 4, 5} & i < 1}`

□

VDM++ contains 13 operators for mappings, the most fundamental of which is the mapping application. This takes a mapping m and a domain value d as its arguments and returns the range value pointed to by d in m. The domain value must really be in the domain of the mapping, otherwise the result of the application is undefined, like indexing out of range in a sequence. Mapping application is written in the same way as function and sequence application:

```
m(d)
```

For example, the expression

```
{<red> |-> 1, <green> |-> 4, <blue> |-> 9} (<blue>)
```

yields the value 9. Suppose mapping m has the value

```
{"Paul" |-> 7, "John" |-> 8, "Nico" |-> 8}
```

then

```
m("Nico")
```

yields the value 8, but the expression

```
m("Marcel")
```

is undefined.

The remaining mapping operators are given in Table 8.1, and each of them is introduced in the larger example in the next section.

Table 8.1: Summary of VDM++ mapping operators

Operator	Name	Type
`dom m`	Domain	`(map A to B)` → `set of A`
`rng m`	Range	`(map A to B)` → `set of B`
`m1 munion m2`	Merge	`(map A to B)*(map A to B)` $\overset{\sim}{\rightarrow}$ `map A to B`
`m1 ++ m2`	Override	`(map A to B)*(map A to B)` → `map A to B`
`merge ms`	Distributed merge	`set of (map A to B)` $\overset{\sim}{\rightarrow}$ `map A to B`
`s <: m`	Domain restrict to	`(set of A)*(map A to B)` → `map A to B`
`s <-: m`	Domain restrict by	`(set of A)*(map A to B)` → `map A to B`
`m :> s`	Range restrict to	`(map A to B)*(set of B)` → `map A to B`
`m :-> s`	Range restrict by	`(map A to B)*(set of B)` → `map A to B`
`m(d)`	Map apply	`(map A to B)*A` $\overset{\sim}{\rightarrow}$ `B`
`m1 = m2`	Equality	`(map A to B)*(map A to B)` → `bool`
`m1 <> m2`	Inequality	`(map A to B)*(map A to B)` → `bool`
`inverse m`	Map inverse	`inmap A to B` → `inmap B to A`

8.3 The Congestion Warning System Revisited

The previous chapter introduced the CWS system and presented a model using sequences to model ordered collection, but the system modelled was rather unrealistic. Consider the following aspects.

- The way the road network was modelled was highly simplified. Adding a new congestion monitor, for example, required inserting a new element in the sequence modelling the road network. Each time this was done, the "location" (index) of each element after the newly inserted one had to be changed. This is not, in itself, a problem, but it makes it difficult, for example, to relate the location of a congestion monitor or sensor to a geographic coordinate.
- There was no notion of *lanes* on the motorway. In reality, multiple sensors can be used at a given location, each gathering the data from vehicles passing over a

different lane. The congestion monitor then combines the data from the different sensors at a location and decides whether congestion exists.

- Having a single actuator manager is not realistic when implementing the system over a wide area. In that case a collection of actuator managers is used, each controlling a number of locations. An actuator manager needs to communicate about the signals being shown in a different area or to send a signal to be shown in the area controlled by another actuator manager. To keep track of which actuator manager controls which area, a *name server* is used.
- In the previous model, congestion monitors, the corresponding sensors and actuators were all added to the system at one time. In practice the infrastructure for the system (sensors and actuators) is added first, then actuator managers and congestion monitors are added.

Figure 8.1 presents an overview of a CWS that deals with these features.

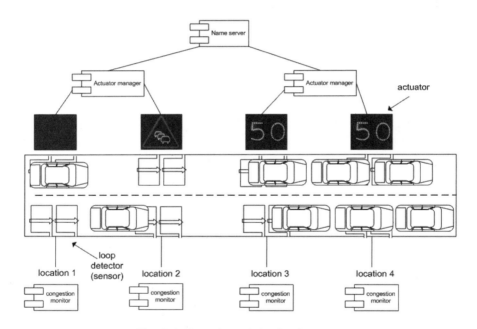

Fig. 8.1: Overview of the CWS system

The first step in modelling this more advanced system is to change the basic data structures defined in class CWS to reflect the new features. How can it be ensured that the location of a component of the system no longer automatically changes when a new component is added? This can be done by regarding the location as a separate entity from the component. The relationship between locations and components could then be modelled using a mapping. For the instance variable roadNetwork this means that it can be modelled as a mapping (initially empty) from locations to congestion monitors:

```
class CWS
...
instance variables
roadNetwork: inmap Location to CongestionMonitor := {|->}
end CWS
```

Notice that the `inmap` type constructor is used because the mapping is injective, at each location only one congestion monitor can exist but at the same time, each congestion monitor can exist at only one location. Defining a mapping as injective provides a powerful mapping operator called `inverse`, introduced later in this chapter.

The definition for the instance variable `sensors` can be adjusted following the same line of reasoning. One more thing to take into account here is the notion of "lanes": Each lane can potentially be associated with a sensor. The definition for `sensors` therefore becomes

```
class CWS
...
instance variables
sensors: inmap Location to (inmap Lane to PassageSensor)
        := {|->};
end CWS
```

The overall structure of the corresponding UML class diagram is given in Figure 8.2. The relation between `CWS` and `CongestionMonitor` is shown as a *qualified association*. An association has been introduced with class `NameServer`, used for controlling access to actuator managers. Note that in the UML class diagram in Figure 7.4 this was shown by means of ordered associations.

A simple operation is introduced to add a new passage sensor to the system at location `loc` and sensing lane `lane`. Effectively, this operation needs to adjust the instance variable `sensors` in such a way that it is updated with the new sensor. The way this can be done depends on whether a sensor has already been placed at location `loc`. It is therefore necessary to test whether `loc` is in the *domain* of `sensors`. For this purpose the VDM++ mapping operator `dom` is available. The domain operator takes a mapping as its argument and yields the *set of values* in its domain. It is written as:

```
dom map expression
```

Suppose that mapping m has the value

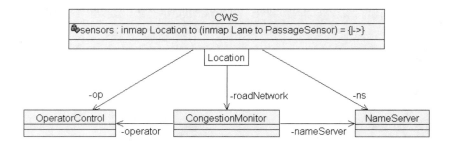

Fig. 8.2: UML class diagram for CWS

```
{"Breakfast" |-> "Banana", "Lunch" |-> "Sandwich",
 "Dinner" |-> "Steak" , "Snack" |-> "Sandwich"}
```

Then the expression

```
dom m
```

will yield the set

```
{"Breakfast", "Lunch", "Dinner", "Snack"}
```

Similarly, a *range operator* exists that does exactly the opposite: Instead of return-ing the set of values in the domain of the mapping, it returns the set of values in its range. It is written as

```
rng map expression
```

Given the definition of mapping m, the result of

```
rng m
```

will be the set

```
{"Banana", "Sandwich", "Steak"}
```

To test whether the mapping `sensors` already uses a location `loc`, a simple set membership test can be used on the domain of `sensors`:

```
if loc in set dom sensors
...
```

Exercise 8.3 What are the results of the following domain and range expressions?

1. dom { |-> }
2. dom {3 |-> <Apple>, 4 |-> <Apple>}
3. rng {3 |-> <Apple>, 4 |-> <Apple>}
4. dom {3 |-> 4} union rng {3 |-> 4}

□

If the location is in the domain of `sensors`, then the range element of `sensors` at `loc` needs to be updated. A *state designator* can be used for this purpose on the left-hand side of an assignment:

```
sensors(loc) := ...
```

The mapping at `sensors(loc)` needs to be updated with the information for the added sensor. Notice that this is a generalisation of the assignment statement introduced in Section 5.7.2. For this purpose the VDM++ *mapping merge operator* can be used. This operator takes two mappings with the same domain and range types as its arguments and returns a *single* mapping consisting of all the maplets in both arguments. The operator is written as an infix operator

```
map1 munion map2
```

If *map1* and *map2* have some domain elements in common, it is still possible to merge them, provided these common domain elements point to the same range elements in both mappings. However, if there is some inconsistency between the two mappings, the result of the merge is undefined. For example, the expression

```
{1 |-> "Apple", 3 |-> "Banana"} munion
{2 |-> "Strawberry", 3 |-> "Banana"}
```

yields

```
{1 |-> "Apple", 2 |-> "Strawberry", 3 |-> "Banana"}
```

but the expression

```
{1 |-> "Apple", 3 |-> "Banana"} munion
{2 |-> "Strawberry", 3 |-> "Cherry"}
```

will yield a run-time error because 3 maps to two different values: `"Banana"` and `"Cherry"`.

In the CWS example, the mapping merge operator is applied as follows. The left-hand side of the operator is a *mapping application* expression representing the current value of `sensors(loc)`. The right-hand side consists of a *mapping enumeration* consisting of one element: the new value that binds the given lane to the newly defined passage sensor:

```
sensors(loc) := sensors(loc) munion
                {lane |-> passageSensor}
```

If the location is *not* in the domain of `sensors`, the entire `sensors` mapping needs to be updated. Again, using the `munion` operator, this can be written as:

```
sensors := sensors munion
           {loc |-> {lane |-> passageSensor}}
```

The complete definition of the operation can be written as:

```
class CWS
...
operations

public AddSensor: Location * Lane ==> ()
AddSensor(loc, lane) ==
def passageSensor = new PassageSensor(loc, lane)
in
  let sensorAtLane = {lane |-> passageSensor}
  in
    if loc in set dom sensors
    then sensors(loc) := sensors(loc) munion sensorAtLane
    else sensors := sensors munion {loc |-> sensorAtLane}
end CWS
```

Exercise 8.4 Specify an operation `AddCongestionMonitor` that takes a location as its argument, creates a new congestion monitor and adds it to `roadNetwork`.

Make use of the `munion` operator. Remember to specify an appropriate precondition.
□

The next class defined is `NameServer`. Whereas in the example in the previous chapter only one actuator manager existed, now a collection of such managers is introduced, each controlling actuators at a number of locations. To keep track of which actuator managers controls which location(s), a name server is introduced. The name server provides several kinds of services that can be used by the actuator managers themselves to find out which actuator manager controls a given location, which are the valid locations, etc.

The class has an instance variable, which models the relationship between the actuator managers available and the locations they control. Because each actuator manager typically controls several locations and their ordering is not relevant, these locations can best be modelled as a set. The relationship between actuator managers and these sets of locations can again best be modelled as a mapping:

```
class NameServer
...
instance variables
am: map ActuatorManager to (set of CWS`Location) := {|->}
end NameServer
```

Exercise 8.5 Formulate an invariant for the state of class `NameServer` expressing that no locations are managed by more than one actuator manager. *Hint:* Use the mapping range operator. □

The first operation defined in this class takes an actuator manager and its associated set of locations as an argument and updates instance variable am to reflect this new information. It is assumed here that the actuator manager may already be present in am and that the new information must *replace* the old information, if present. For this purpose the VDM++ *mapping override* operator can be used.

The mapping override operator takes two mappings *map1* and *map2* as its arguments and yields the merge of the two mappings, except that where there is a conflict, *map2* overrides *map1*. This is written as

```
map1 ++ map2
```

For example, the expression

```
{"Peter" |-> <vegetarian>, "Nico" |-> <meat>} ++
{"Peter" |-> <meat>, "John" |-> <vegetarian>}
```

yields

```
{"Peter" |-> <meat>,
 "Nico"  |-> <meat>,
 "John"  |-> <vegetarian>}
```

Unlike munion, this operator is total: It yields a defined result for any two mappings. Notice also that the order of the arguments matters: "m1 ++ m2" is not the same as "m2 ++ m1".

Exercise 8.6 Calculate the following expressions:

1. {1 |-> 5, 5 |-> 1} munion {3 |-> 1, 1 |-> 5}
2. {1 |-> 5, 5 |-> 1} ++ {3 |-> 1, 1 |-> 5}
3. {1 |-> 5, 5 |-> 1} munion {3 |-> 1, 1 |-> 6}
4. {1 |-> 5, 5 |-> 1} ++ {3 |-> 1, 1 |-> 6}

□

The corresponding operation in class NameServer can thus be written as:

```
class NameServer
...
operations

public SetActuatorManager: ActuatorManager *
                            set of CWS'Location ==> ()
SetActuatorManager(actuatorManager, locations) ==
  am := am ++ {actuatorManager |-> locations};
end NameServer
```

It is important that the name server can identify which actuator manager controls a given location. The information on which locations are controlled by each actuator manager is available in the instance variable am. However, instead of having an actuator manager and using that as an "index" in am to retrieve the associated locations, here it needs to be done the other way around, having a (set of) locations and using that as an index to retrieve the associated actuator manager. Because the mapping application can only be done using a domain element, the mapping here needs to be "inverted." For this purpose, VDM++ has an *inverse* operator.

The inverse operator takes an injective mapping and yields a mapping in which all domain values have been replaced by the corresponding range values and vice versa. It is written as:

```
inverse map
```

For example,

```
inverse {"Peter" |-> <vegetarian>, "Nico" |-> <meat>}
```

yields

```
{<vegetarian> |-> "Peter", <meat> |-> "Nico"}
```

Exercise 8.7 Why is the inverse operator only defined for injective (inmap) mappings? □

The corresponding operation in class NameServer can now be written. For reasons that will become clear when the specification of class ActuatorManager is discussed later in this chapter, both the type of the input parameter and the type of the output parameter are defined as *optional*; if a nil value is passed as an input parameter then the operation will return the nil value as a result as well:

```
class NameServer
...
public
GetActuatorManager: [CWS`Location] ==> [ActuatorManager]
GetActuatorManager(loc) ==
  if loc = nil
  then return nil
  else let locations = inverse am
       in
          let locationSet in set dom locations be st
              loc in set locationSet
          in
             return locations (locationSet);
end NameServer
```

Exercise 8.8 Write an operation GetLocations, that takes no arguments and returns the *set* with all controlled locations by the name server. □

8.3.1 Sensors and Traffic Controls

Consider now the hierarchy of sensors and traffic controls. Only a few changes have to be made over the sequence-based example presented in the last chapter. In fact, the only changes in this part of the example are made in class CongestionSensor. Whereas it was previously only possible to associate a single passage sensor with a congestion sensor, it is now possible to have multiple passage sensors associated with a single congestion sensor: one for each lane at the location of the congestion sensor. Again, this notion can be modelled as a mapping:

```
class CongestionSensor
...
instance variables
messageLog: map CWS'Location to (seq of (seq1 of char))
          := {|->}
end CongestionSensor
```

Exercise 8.9 Would it be possible to use a *set* of passage sensors here instead of a *mapping* associating locations with passage sensors? □

The operation that needs to be modified next is `IssueCongestionStatus`. The algorithm used to calculate the congestion status involves finding the minimum of all the average speeds obtained from the available passage sensors. This is then compared with the defined threshold values to determine congestion.

The first step of this calculation is an iteration over all the available lanes in which a passage sensor is present. These are the lanes available in the domain of the mapping `passageSensors`; the `dom` operator can be used here. The associated average speeds are then gathered in a set of speed values using a set comprehension expression:

```
{passageSensors(lane).AverageSpeed(NoPassages)
 | lane in set dom passageSensors}
```

The minimum of these values is calculated by a function `min` that returns the smallest of a set of numeric values. A *function* is used here because of its generic nature: calculation of the minimum of a set of values is something that is likely to be used in other examples as well. By having a function for this, chances are that it can be reused and efforts can be saved during other specification activities.

The concept behind the function `min` is quite simple. The function is based on the observation that the minimum value of a set of natural numbers is *any* value from that set, *provided* there are no other numbers in that set smaller than the chosen number. The use of the VDM++ *let-be-such-that expression* is very well suited for this. The function `min` can be written as:

```
class CongestionSensor
...
functions
min: set of nat -> nat
min(numbers) ==
```

```
   let n in set numbers be st
     not exists n2 in set numbers & n2 < n
   in
     n
end CongestionSensor
```

and it is then possible to give the final version of IssueCongestionStatus:

```
class CongestionSensor
...
public IssueCongestionStatus:
          () ==> CongestionSensor'CongestionStatus
IssueCongestionStatus() ==
  def averagespeed = min ({passageSensors(lane).
                                  AverageSpeed(NoPassages)
                             | lane in set
                               dom passageSensors})
  in
     if averagespeed < CongestionThreshold
     then return <Congestion>
     elseif averagespeed > NoCongestionThreshold
     then return <NoCongestion>
     else return <Doubt>
end CongestionSensor
```

8.3.2 Actuators

Significant changes need to be made to class ActuatorManager to reflect the new requirements. Starting with the instance variable section, a name server has to be added because an actuator may need to consult other actuator managers if any of its operations require access to the actuators outside its own area. The definition of instance variable as needs to be changed to reflect the more flexible structure relating locations to actuators. For the latter purpose, again, a mapping is appropriate:

```
class ActuatorManager
...
instance variables
as: inmap CWS'Location to Actuator := {|->};
ns: NameServer

end ActuatorManager
```

One of the shortcomings that was detected in the previous version of the example was that there are no facilities for adding or deleting individual components (sensors, actuators, etc.) from the system. This is solved in this version by defining operations to add, remove and replace actuators.

The first operation defined is `AddActuator`, which takes a location as its argument, creates a new actuator and adds the resulting maplet to its collection of controlled actuators. For this purpose the *distributed mapping merge* operator can be used. The distributed mapping merge operator takes a set of mappings `sm` as its argument and returns a mapping that contains the union of all maplets in the mappings that are a member of `sm`. It is written as

```
merge sm
```

For example, the expression:

```
merge {{<France> |-> 9, <Spain> |-> 4},
       {<France> |-> 9, <England> |-> 3, <USA> |-> 1}}
```

yields

```
{<France> |-> 9, <Spain> |-> 4,
 <England> |-> 3, <USA> |-> 1}
```

The distributed mapping merge operator can be used in the definition for the operation `AddActuator` as follows:

```
as := merge {as, {loc |-> actuator}}
```

Another issue that needs to be resolved here is that the name server must be informed that the actuator manager controls a new actuator at the given location. Class `NameServer` provides an operation `SetLocation` for this purpose, which takes an actuator manager and a location as its argument. In this case, the actuator manager itself reports that it controls a new location; the call to this operation must pass an object reference to that actuator manager as an argument. This, in VDM++ is written as `self`, and the call to the operation is written as:

```
ns.SetLocation(self, loc)
```

The specification for `AddActuator` can be written as follows:

```
class ActuatorManager
...
operations

public AddActuator: CWS`Location ==> ()
AddActuator(loc) ==
  def actuator = new Actuator()
  in
  (as := merge {as, {loc |-> actuator}};
   ns.SetLocation(self, loc)
  )
pre loc not in set dom as;

public GetSignal: [CWS`Location] ==>
                   [CongestionMonitor`Signal]
GetSignal(location) ==
  if location = nil or location not in set dom as
  then return nil
  else return as(location).GetSignal();
end ActuatorManager
```

Exercise 8.10 Redefine the operation AddActuator by using a *mapping merge operator* instead. □

A second operation, called RemoveActuator arranges for an actuator to be *deleted* from the collection of actuators controlled by the actuator manager. Deletion of maplets from a mapping is, however, not possible given the mapping operators discussed so far, and in fact what is needed here is a mechanism that is capable of filtering mappings based on given criteria. For this purpose four different operators are defined in VDM++:

- The *domain-restrict-to* operator ("< :"): This takes a set s as its left-hand operand and a mapping m as its right-hand operand. It returns a mapping that contains all maplets in m whose domain element is also a member of s. It is not actually required that the elements in set s are also in the domain of m.
- The *domain-restrict-by* operator ("<-:"): This takes a set s as its left-hand operand and a mapping m as its right-hand operand. It returns a mapping that contains all maplets in m whose domain element is *not* a member of set s.
- The *range-restrict-to* operator (":>"): This takes a mapping m as its left-hand operand and a set s as its right-hand operand. It returns a mapping that contains all maplets in m whose range element is also a member of set s. It is not required that the elements in s are also in the range of m.

- The *range-restrict-by* operator ("`: ->`"): This takes a mapping m as its left-hand operand and a set s as its right-hand operand. It returns a mapping that contains all maplets in m whose range element is *not* a member of s.

It is worth considering a few examples using these operators. Suppose that mapping m has the value

```
{<France>  |-> 9, <Denmark>  |-> 4,
 <SouthAfrica>  |-> 2, <SaudiArabia>  |-> 1}
```

and that set s has the value

```
{<France>, <England>, <Denmark>, <Spain>}
```

Then s `<:` m yields the value

```
{<France>  |-> 9, <Denmark>  |-> 4}
```

The expression s `<-:` m yields the value

```
{<SouthAfrica>  |-> 2, <SaudiArabia>  |-> 1}
```

The expression m `:>` {2,..., 10} yields the value

```
{<France>  |-> 9, <Denmark>  |-> 4, <SouthAfrica>  |-> 2}
```

and the expression m `:->` {2,..., 10} yields the value

```
{<SaudiArabia>  |-> 1}
```

Exercise 8.11 Calculate the following expressions:

1. {<b>, <d>} `<-:` {<a> |-> 1, <c> |-> 3, <d> |-> 1}
2. {<b>, <d>} `<:` {<a> |-> 1, <c> |-> 3, <d> |-> 1}
3. {<a> |-> 1, <c> |-> 3, <d> |-> 1} `:->` {1, 3}
4. {<a> |-> 1, <c> |-> 3, <d> |-> 1} `:>` {1, 3}

□

This powerful mechanism can easily be used for the definition of the operation RemoveActuator. What is needed here is that the range of the (mapping) instance variable as is restricted by the parameter (actuator) passed to the operation. This leads to the following definition:

```
class ActuatorManager
...
operations

public RemoveActuator: Actuator ==> ()
RemoveActuator(actuator) ==
  as := as :-> {actuator};
end ActuatorManager
```

Exercise 8.12 The current definition of the operation RemoveActuator does *not* remove the location from the collection of locations controlled by the actuator manager at the name server. Adjust the operation to achieve this and define an operation in class NameServer for this purpose. □

For the definition of the operation ReplaceActuator, two operands are used, which were introduced earlier. These are the *inverse* and *override* operators. The inverse operator is needed to define at which location an actuator needs to be replaced. The override operator is then used to replace the old actuator with the new one. This results in the following definition for the operation:

```
class ActuatorManager
...
operations

public ReplaceActuator: Actuator * Actuator ==> ()
ReplaceActuator(actuator, newActuator) ==
  as := as ++ {(inverse as)(actuator) |-> newActuator};
end ActuatorManager
```

Exercise 8.13⋆ Rewrite the operation ReplaceActuator twice:

- the first time using the range-restrict-to operator and
- the second time using a let-be-such-that-statement.

□

The most complicated operation in class ActuatorManager is ShowSignal. It has to take into account that information is (possibly) needed on the signal being shown downstream to determine which signal is being shown at the current location (for this, another actuator manager may be needed, information on which can be retrieved via the name server) and that the signal being shown upstream may have to be adjusted (again, the name server may be needed to achieve this, depending on whether the location upstream is under the control of the current actuator manager). In the following definition, notice the use of the *def statement* to gather the information needed for the calculation of the actual signals being shown up front:

```
class ActuatorManager
...
public ShowSignal: CWS'Location *
                   CongestionMonitor'Signal
                   ==> ()
ShowSignal(location, signal) ==
( def downstreamLocation =
        Downstream(location, ns.GetLocations());
      downstreamManager =
        ns.GetActuatorManager(downstreamLocation);
      downstreamSignal =
        if downstreamManager <> nil
        then downstreamManager.
              GetSignal(downstreamLocation)
        else nil;
      actuator = as(location);
      upstreamLocation = Upstream(location,
                               ns.GetLocations());
      upstreamManager = ns.GetActuatorManager(
                          upstreamLocation);
      upstreamSignal =
        if upstreamManager <> nil
        then upstreamManager.GetSignal(upstreamLocation)
        else nil
  in

  ( -- Set the right signal at the current location
    ShowSignalAtLoc(signal,downstreamLocation,
                    downstreamSignal,actuator);

    -- Set the right signal upstream
    ShowSignalUpstream(signal,upstreamLocation,
                      upstreamManager,upstreamSignal)
  )
)
pre location in set dom as;
end ActuatorManager
```

8.3.3 Operator Control

The final class that is revisited in the CWS example is `OperatorControl`. To ease manipulation of log messages and to make access to the log messages possible on the basis of the location to which they apply, a mapping is used that relates locations to sequences of messages that were issued for that location:

```
class OperatorControl
...
instance variables
messageLog: map CWS'Location to (seq of (seq1 of char))
         := {|->}
end OperatorControl
```

The operation `WriteLog`, which takes a new message applicable to a given location and appends that message to the ones already collected, first checks whether messages for that location have already been gathered and then uses the mapping override operator to adjust `messageLog` accordingly:

```
class OperatorControl
...
operations

public WriteLog: seq1 of char * CWS'Location ==> ()
WriteLog(message, location) ==
  let newMessage = message ^ int2String(location),
      messages = if location in set dom messageLog
                 then messageLog(location) ^
                      [ newMessage ]
                 else [ newMessage ]
  in
    messageLog := messageLog ++ {location |-> messages};
end OperatorControl
```

Finally, the operation that returns information on all locations at which congestion has occurred at some time can easily be modelled using the mapping domain operator :

```
class OperatorControl
...
operations

public CongestionSpots: () ==> set of CWS'Location
```

```
CongestionSpots() ==
  return dom messageLog;
end OperatorControl
```

8.4 Summary

This chapter has presented the mapping type constructor. Mappings are unordered collections of values (maplets) that make it possible to associate a domain value to a range value. As such they provide a very powerful and flexible data abstraction. The various operators for mappings have been described and demonstrated in this chapter. These operators allow convenient manipulation of mapping values, allowing the modeller to concentrate on the salient features of the problem at hand. The extended CWS example demonstrates the value of mappings for describing general and hierarchical structures based on name spaces.

Part III

Modelling in Practice: Three Case Studies

9

Model Structuring: The Enigma Cipher

9.1 Introduction

The elements of object-oriented modelling in VDM++ have now been introduced and the focus can move from theory to practice. Many questions about the use of this technology will already have occurred to the reader. What do "real" models look like? How does modelling relate to testing? How does one use the formal models of data and function (in VDM++) alongside class diagrams in UML? And what are the effects of applying this on projects?

The next three chapters aim to address these questions by means of three case studies. The key theme is that of structuring as a way of dealing with complexity and scale. The study of the Enigma cipher machine in this chapter shows how the object-oriented structure of a model reflects that of the problem it is used to analyse and that this pays off in the ability to take a structured approach to testing. Chapter 10 discusses in greater depth the way UML class diagram models work with VDM++ descriptions of data and functionality. Chapter 11 brings these themes together by telling the story of a commercial application in which UML models derived from an enterprise architecture combine with detailed VDM++ models of data and functionality, and with systematic testing, to yield benefits in terms of product quality and development cost.

This chapter shows the development of a VDM++ model of the famous Enigma cipher machine used by the Germans in the Second World War to encrypt and decrypt messages that were exchanged between military units. The purpose of the model is to get a basic understanding of the cipher mechanism as implemented in Enigma. It is often difficult to apply a new technique for the first time, so the model will be presented in some detail, with a careful explanation of each modelling step and the rationale of many of the design decisions taken during its construction. First (Section 9.4) a UML model will be suggested, revised and restructured before the data and function models in VDM++ are presented in greater detail. After building the model, its validity will be analysed using structured testing techniques in Section 9.5.

Simon Singh's popular scientific work *The Code Book* ([Singh99]), Robert Harris's novel *Enigma* ([Harris95]) and the related major motion picture have certainly raised interest in Enigma recently. In spite of Enigma's fame, there are relatively few

attempts at modelling the machine in the literature, making it an intriguing example
for this chapter.

9.2 The Historic Significance of Enigma

The Enigma cipher machine, Enigma for short, played a significant role in modern
history. The outcome of the Second World War was certainly influenced by the capabilities of the Allies to crack Enigma (and later Lorentz) ciphers at Bletchley Park,
where special machines were designed to automate the code-breaking process. The
first generation of those machines were christened *Bombe* and they were built using
electromagnetic relays. Anecdotal evidence says that the name was inspired by the
ticking noise the relays made during its operation. The first Bombe (called *Victory*)
was installed in March 1940 at Bletchley Park, and it was used to crack Enigma-
encoded messages. A few months later, some 50 machines were working in parallel to
process all intercepted messages.

With the advent of the much more complex Lorentz ciphers later in the war, the
capabilities of the Bombe were deemed insufficient and the mathematicians and engineers at Bletchley Park designed a new system based on electronic valves, which
was cutting-edge technology at the time. The project was headed by Max Newman,
and the chief consultant was mathematician Alan Turing, who also played a major
role in designing the Bombe. No fewer than 1500 vacuum tubes were needed to build
it, giving rise to its name: *Colossus*. The Colossus was installed at Bletchley Park in
December 1943, and it could crack on average 300 messages per month. The number
of intercepted Lorentz messages nevertheless increased substantially and the performance of Colossus needed to be improved. A new project was set up to deliver the
Colossus Mark-II on June 1, 1944 – only five days before D-Day. The Colossus Mark-
II was five times faster than its predecessor. It consisted of 2500 valves and could be
programmed; it was one of the first programmable and electronic computers.

After the war, the British government in particular did not wish to disclose its code-
breaking capabilities. This is the main reason why ENIAC is generally considered to
be the world's first electronic computer, but in fact Colossus was operational about a
year earlier. Alan Turing developed most of his theoretical insights in computing at
Cambridge in the years before the war. His work on undecidability and the universal
Turing machine are now famous. At Bletchley Park he succeeded in applying these
theoretical results to break the Enigma code. When security restrictions were lifted in
the early 1970s and the significance of the results from Bletchley Park was realised,
credit was finally given to those who had earned it. Turing is now widely acclaimed
as the founding father of computing science, but he never personally enjoyed that
status. Historians believe that he committed suicide in 1954 because he was tried,
convicted and openly humiliated for being homosexual. Andrew Hodges's biography
Alan Turing: The Enigma ([Hodges92]; see also http://www.turing.org.uk)
provides more insight into the life and work of this brilliant scientist.

9.3 The Enigma Cipher Algorithm

The Enigma cipher machine was designed and developed by Arthur Scherbius and Richard Ritter in 1918. These German inventors created an electronic equivalent of the cipher disk invented by Leon Alberti around 1400. The cipher disk is a simple device that substitutes letters. The earliest substitution cipher is known as the Caesar shift cipher (or Caesar cipher), where each letter is simply replaced by a letter that is a certain distance further along in the alphabet. For example (with an encoding distance of 4) A will be represented by E, B by F, C by G and so on.

The cipher disk of Alberti can be used for the encoding and decoding of messages. The cipher disk actually contains two disks, an inner and an outer disk, both containing 26 positions, each representing a letter in the alphabet. By rotating the inner disk against the outer disk the encoding distance can be set. If A on the inner ring is set adjacent to E on the outer ring, then all other letters will be automatically aligned properly. The encoding (and decoding) of a message is now a simple task that can be performed, for example, directly on the battlefield.

Exercise 9.1 Consider a monoalphabetic cipher such as the Caesar cipher. What is the easiest way to prevent the opponent from guessing the encoding distance? What is the principle weakness of monoalphabetic ciphers? □

The encryption is obviously very weak; you simply have to guess the encoding distance, and there are only 25 possibilities. This is typical for so-called monoalphabetic ciphers. However, Alberti used the same device to make code breaking much more difficult. He used a code word to implement the encoding distance. Take, for example, his own first name, "LEON." Instead of taking a fixed distance, the encoding distance is now determined by the order of the letters in the code word. To encode the first letter, the inner disk is set such that the letter L on the inner ring faces A on the outer ring. Encoding the letter N will now yield C. To encode the second letter, the inner disk is reset such that E (the second letter of the code word) faces A. Encoding the letter H will now yield D. This process is repeated for each letter; the fifth letter is again encoded with the first letter of the code word and so on. This so-called polyalphabetic cipher is much harder to break because for every letter another encoding distance is selected. This technique is also known as the Vigenère cipher. An elegant and abstract specification of a polyalphabetic cypher in VDM–SL can be found in [Jones90] (pages 179–81).

These basic techniques were actually used to design the Enigma. Scherbius and Ritter combined several strategies to improve the strength of the cipher, creating a device that was robust, reliable and very easy to use. The cryptographic power of the Enigma is in sharp contrast with the simple means needed to build such an ingenious device. Scherbius was awarded a patent on Enigma in 1918, but still it took him several years to sell it to the German military. Eventually some 30,000 devices were ordered by the German armed forces. Most notable was the use of Enigma in the German navy where Admiral Dönitz, commander of the U-boat fleet, ordered a special version with even greater cipher strength than the standard Wehrmacht Enigma.

The Enigma cipher machine is basically a typewriter composed of three parts: the keyboard to enter the plain text, the encryption device and a display to show the cipher text. Both the keyboard and the display consist of 26 elements, one for each letter in the alphabet. White space in the plain text was simply ignored, which incidentally makes code breaking even more difficult because word boundaries are removed. The operator, typically a signals person, such as a Marconist aboard a ship, configured the Enigma system once per day and encoded the message by typing each letter and writing down the letter highlighted on the display. The encoded message was then sent using normal Morse code by radio. Of course, the enemy could also receive the Morse signal, but they could not understand its contents without cracking the Enigma code. The intended receiver could simply decode the message by repeating the encoding process on the cipher text. Enigma was configured in the same way as the sender had and the message was decoded by typing in each encoded letter and reading the decoded letter from the display. Having the same procedure for encoding and decoding made the machine very easy to use; errors were seldom made.

The encryption device in the Enigma consists again of three parts: a plugboard, a set of rotors and the reflector. The plugboard is used to create a fixed mapping in which each letter can be *replaced* by any other letter. The board is configured by manually inserting patch cables. The most important part of the Enigma is the so-called rotor, also called the scrambler disk or scrambler. The standard Enigma had three rotor slots (places where rotors could be inserted in the system) and five scrambler disks; the naval version had four rotor slots and eight scrambler disks. The scrambler disks were universal and could be inserted in any Enigma rotor slot, creating a vast number of possible Enigma configurations. A scrambler disk actually consists of two disks with 26 electrical contacts each. Every contact point on disk 1 is electrically wired to a specific contact point on disk 2, creating a fixed mapping. The rotor slots were organised such that the contact points on disk 2 touched the contact points of disk 1 on the next scrambler disk. In this way, electrical current could flow from the plugboard, through all rotor disks toward the reflector. The cryptographic strength of Enigma was increased further by turning the scrambler disks, so that each time a different mapping is used when a letter is encoded (or decoded).

The ingenuity in the design is particularly evident in the way the rotors operate. The right-most rotor, which is closest to the plugboard, is turned whenever a key is pressed on the keyboard. Each rotor has a latch at one of the 26 positions, for each rotor a different position. If the latch is adjacent to the latch of the disk on the left, that rotor will also proceed to the next position (just like the old analog odometer in a car). Whether a rotor moves to the next position at each keystroke is dependent on which rotor is in which slot, the position of the latch on each rotor and the start positions of the rotors.

The final piece of the encryption device is the reflector. It is actually responsible for the symmetric (reciprocal) nature of Enigma; encoding and decoding are identical processes. The reflector is placed at the end of the rotor chain. The principal difference between the rotor and the reflector is that the reflector consists of only one disk with 26 electrical contacts, which are all interconnected in pairs. This means that the electric current coming from the last scrambler disk is fed back to that same disk but on a different position. So the electronic current flows twice through the set of scrambler disks before it reaches the display (seven substitutions for standard Enigma, nine for the naval version). The standard Enigma has a fixed-position reflector, the naval version has an adjustable reflector; it can be set in 26 positions, again to increase the cipher strength.

Enigma was deemed impenetrable. The code was eventually broken due to a combination of luck, weaknesses in the Enigma design (in particular the reflector), wrong use of the code words by the Germans and the endurance and brilliant analysis of the mathematicians and linguists at Bletchley Park. The historic relevance of the Enigma and the people trying to break it can be investigated further, for example, at `http://www.bletchleypark.co.uk`, which also has an on line Enigma simulator. Other Enigma resources are `http://www.bletchleypark.net` and the Web sites of the Deutsches Museum at Bonn (Germany) and the National Cryptologic Museum near Washington, D.C. (USA). Last, but not least, the Web site of Tony Sale: `http://www.codesandciphers.org.uk`. He is the curator of the Bletchley Park Museum and his Web site provides an excellent overview of the Enigma and the machines that were developed to break its code.

9.4 Building the Enigma Model

The information provided in the previous section is sufficient to start modeling Enigma using VDM++. The approach proposed in Chapter 2 will be used by applying the guidelines to the problem. The first guideline is to determine the *purpose of the model*. It is important to gain a *conceptual understanding* of the Enigma device and in particular to investigate the *encryption algorithm* used. The model should reflect the structure of the Enigma device, so that it is possible to investigate the behaviour of each component in isolation as well as the behaviour of a (sub)system that is composed of several components. VDM++ is well suited for describing these kind of architectures.

Interaction with the model is essential to reach these goals. Syntax and type checking can be used to increase confidence in the internal consistency of the model, while prototyping and testing are techniques to validate that the model actually reflects reality. Therefore, as a secondary aim, *structured testing* techniques are presented to show how confidence in the model can be increased. Bearing in mind the purpose of the model, it is not appropriate to present a minutely detailed replica of Enigma; the aim is really to better understand how the Enigma cipher works and to see how this can be modelled effectively in VDM++.

The second step is to create a dictionary of candidate classes, data types and operations that can be extracted from the information provided in the previous section. This step actually covers guidelines 2, 3 and 4 presented in Chapter 2. Such a dictionary could appear as follows:

Potential Classes and Types (Nouns)

- *Reflector: configuration (which input on the disk is connected to which output on the same disk), current position (which input is at absolute position 1 when a character is substituted)*
- *Rotor: configuration (which input on disk 1 is connected to which output on disk 2), current position (which input is at absolute position 1 when a character is substituted), latch position (at what input is the latch located on the rotor, in order to propagate the rotation to the adjacent rotor when the latches are at the same absolute position), next_rotor (the next rotor in the chain), reflector (the reflector connected to the last rotor in the chain)*
- *Plugboard: configuration (which character is replaced by which other character and vice versa; substitutions are always in pairs), first_rotor (which rotor is connected to the plugboard)*
- *Enigma: plugboard (top-level component of the system containing a link to the plugboard)*

Potential Operations (Actions)

- *Keystroke: characters are encoded (or decoded) one by one*
- *Substitute: transposition of a character to another character in the alphabet*
- *Encode & Decode: the transposition of a character is dependent on the direction of the information flow (to or from the reflector, respectively)*
- *Rotate: the rotor turns one position, e.g., after each keystroke or when the latch on the rotor is adjacent to the previous rotor in the chain*

An initial model based on the dictionary is constructed following steps 5 and 6 of the guidelines. The result is the UML class diagram shown in Figure 9.1. The corresponding VDM++ specification is not provided here (it can be found on the book's Web site) but the internal consistency of the VDM++ model has been checked using VDMTools. Note that in this initial design, the Keystroke operation accepts and returns characters whereas all Encode, Decode and Substitute operations use natural numbers in their signatures. These numbers represent the position of the character in the alphabet according to some arbitrary ordering scheme (for example "A".."Z"). The transposition relation can be expressed as a mathematical operation on these indices, which is easier than dealing with the individual characters themselves.

All the elements identified in the dictionary seem to be covered in the initial design, but there is some overlap in functionality and data within the proposed classes. This initial design is far from optimal; it needs modification such that a better balance between structure (organisation of the model, also called *architecture*) and functionality is achieved. This is probably the most difficult part of any design process; sometimes it can take several attempts to "get it right." Design patterns (see [Gamma&95]) are often used to structure the model, but deciding which pattern to use remains an experience-based task. It is nevertheless a very important step because wrongly organized models can cause tremendous maintenance problems later in the system life cycle. Also, testability is greatly influenced by the proper organisation of the model. Analysis of the initial model shows that:

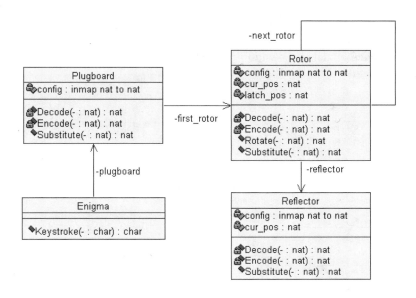

Fig. 9.1: The UML class diagram of the Enigma device – initial design

1. A typical pipe-and-filter architecture (see [Shaw&96]) occurs in the design. Several components are linked in a chain, where the output of a component is used as the input of the next component in the chain. Instead of giving each component a specific relation to the next component in the chain (in the initial design these are first_rotor, next_rotor and reflector), a new class, called Component, is provided that captures this notion of chaining through a single instance variable next.

2. The Plugboard, Rotor and Reflector classes all contain the operations Encode, Decode and Substitute. Some of these functions are identical for each component, while some have slightly adapted behaviour. The instance variable config is also needed in all three classes. It is used to store the way the mapping of indices should occur in each component, but again this is done slightly differently in each case. The instance variable config and the operations Encode, Decode and Substitute are ideal candidates for abstraction by inheritance. Therefore, a new class called Configuration is introduced that captures this abstraction. This functionality is kept separate from Component for two reasons: (1) The top-level class Enigma is potentially a subclass of Component (to point to the plugboard) but it does not need the configuration information and associated functionality contained in the class Configuration and (2) in general, it is wise to keep the structure and functionality separate from each other, for example, to allow reuse of functionality in a different model structure.

The initial design has been modified to achieve a better balance between functionality and structure. Functionality has been grouped so that it is easier to maintain the model because changes do not propagate outside the class boundaries. An overview of the improved model is shown in Figure 9.2. The revised design will be presented one class at a time using the VDM++ notation.

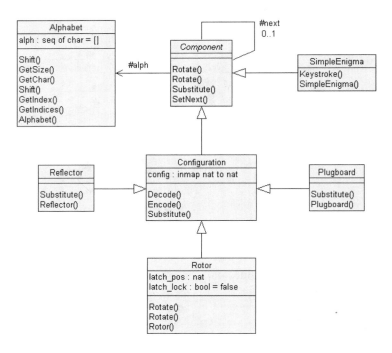

Fig. 9.2: The revised Enigma UML model

9.4.1 Alphabet

The auxiliary class Alphabet gathers all the properties and functionality to describe and manipulate the Enigma alphabet. This is important because otherwise this knowledge would be spread around in the rest of the specification, with bits and pieces everywhere. The chances are high that, for example, functionality would be duplicated and the model would be very difficult to maintain.

The alphabet consists of a sequence of characters. The invariant (as required by step 7 from the guidelines) states that the alphabet must have even length and all characters in the sequence must be unique. The even length property is needed by the Reflector. Each input on the disk should be connected to exactly one output and vice versa. This requirement can be met only if an even number of contact points, and thus an even length alphabet, is used.

```
class Alphabet

instance variables
  alph : seq of char := [];

inv AlphabetInv(alph)

functions
  AlphabetInv: seq of char -> bool
  AlphabetInv (palph) ==
    len palph mod 2 = 0 and
    card elems palph = len palph
```

The Alphabet class provides a constructor for creating a new Alphabet instance and operations GetChar to retrieve the nth character from the alphabet, GetIndex to retrieve the index of the character in the sequence, GetIndices to return the set of all possible indices, and GetSize to return the size of the alphabet. These are shown here.

```
operations
  public Alphabet: seq of char ==> Alphabet
  Alphabet (pa) == alph := pa
  pre AlphabetInv(pa);

  public GetChar: nat ==> char
  GetChar (pidx) == return alph(pidx)
  pre pidx in set inds alph;

  public GetIndex: char ==> nat
  GetIndex (pch) ==
    let pidx in set {i | i in set inds alph
                         & alph(i) = pch} in
      return pidx
  pre pch in set elems alph;

  public GetIndices: () ==> set of nat
  GetIndices () == return inds alph;

  public GetSize: () ==> nat
  GetSize () == return len alph;
```

Note that the set comprehension in GetIndex is always a singleton set (a set containing just one element) due to the uniqueness property stated in the invariant. Therefore, pidx will always be bound to the proper value by the in set operator in the *let* statement.

The substitution algorithm mentioned earlier is the core of Enigma encryption and decryption. It relies on the ability to transpose characters in the alphabet. These transpositions are expressed by the overloaded Shift operations, which calculate a new index based on the current index and a transposition distance or offset. The second Shift operator is defined for convenience, to transpose a character by just one position. It is used to implement the movement of the Rotor, which will be presented later:

```
public Shift: nat * nat ==> nat
Shift (pidx, poffset) ==
   if pidx + poffset > len alph
   then return pidx + poffset - len alph
   else return pidx + poffset
pre pidx in set inds alph and
     poffset <= len alph;

public Shift: nat ==> nat
Shift (pidx) == Shift(pidx, 1)
end Alphabet
```

The strength of the Alphabet class is that it is possible to express an alphabet of arbitrary length and arbitrary ordering, which provides a lot of freedom to experiment with the Enigma model using different alphabets.

9.4.2 Component

The class Component allows construction of a linked list using an instance variable next to point to the next component in the list. An operation SetNext is used to instantiate the link. The instance variable next should not be assigned more than once. Furthermore, loops in the linked list have to be prevented. This can be done in a precondition on the SetNext operation (again step 7 from the guidelines), as shown here. Typically for models of such data structures, recursion is used to gather components already linked into the structure:

```
class Component

instance variables
  protected next : [Component] := nil;
  protected alph : Alphabet
```

```
operations
  public Successors: () ==> set of Component
  Successors () ==
    if next = nil
    then return {self}
    else return {self} union next.Successors();

  public SetNext: Component ==> ()
  SetNext (pcom) == next := pcom
    pre next = nil and
        self not in set pcom.Successors();
```

Exercise 9.2 Write an alternative specification for Successors that does not use recursion. □

The precondition on SetNext prevents loops by testing the self object reference against the set of object references retrieved by Successors, which uses recursion to compute the set of object references of all components that are already in the linked list. Note that the instance variable next is declared protected, which means that derived classes can access it without an extra operation. This modelling style is used here because the model is only instantiated once (at startup) and is not modified during the remainder of its lifetime. Furthermore, each Component has a reference to the Alphabet. The instance variable alph needs to be initialized by the derived classes of Component.

Exercise 9.3 The operation SetNext is defined such that it is not possible to create circular structures, but is the proposed solution here really sufficient? How can it be broken, and what is needed to fix it? □

Each Component should be able to perform two basic tasks: Substitute and Rotate. The Substitute operation takes an index, performs the substitution and finally returns the new index. The implementation of this operation is explicitly delegated to the derived classes of Component by means of the is subclass responsibility construct. The class Component is a so-called *abstract base class* because it is not possible to create an instance of Component directly, due to the delegated operation Substitute.

In contrast, the Rotate operations are executable; their default behaviour is to do nothing. Operation overloading will be used to redefine the behaviour for rotating components (in our case the Rotor) in the derived classes. The first Rotate operation performs immediate rotation, while the second operation takes into account the current position of the latch, which is passed as a parameter. The latter operation is used to implement the latching mechanism of the Rotor, which is presented later.

The reason for using is subclass responsibility in the case of the operation Substitute and skip in the case of the operation Rotate is that knowledge is lacking at this level in the model to say anything useful about the substitution. In contrast, it is possible to be explicit about the components that do not need to rotate. Nevertheless, having both operations defined here allows us to use the statements `next.Substitute()` and `next.Rotate()` anywhere in the model without the need to know the actual subtype of the object reference stored in the instance variable `next`:

```
public Substitute: nat ==> nat
Substitute (-) == is subclass responsibility;

public Rotate: () ==> ()
Rotate () == skip;

public Rotate: nat ==> ()
Rotate (-) == skip

end Component
```

9.4.3 Configuration

The class `Configuration` is generic. It contains an instance variable `config`, an injective mapping between natural numbers, used to represent the transposition relation in Enigma components. In the rotor, this instance variable will describe which input on disk 1 (a number in the domain of the mapping) is connected to which output on disk 2 (a number in the range of the mapping). In the plugboard, the numbers in the domain and range represent the indices of the characters that are swapped. In the reflector, the numbers describe which input (domain) is connected to which output (range) on the same disk. The differences in the way the mapping is used make it very hard to write a useful invariant at the generic class level. In cases like these, it is better to postpone the definition of an invariant to the derived classes where specific invariants can be given. For that reason, the variable is declared `protected` so that derived classes can refer to it directly:

```
class Configuration
  is subclass of Component

instance variables
  protected config: inmap nat to nat;
```

Now that the transposition relation is defined, it is possible to give a generic definition for the `Substitute` operation. Two auxiliary operations, `Encode` and `Decode`, are defined for this purpose. The operation `Encode` implements the encod-

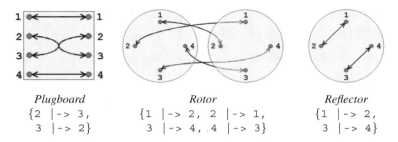

Fig. 9.3: Example configurations for a four-character Enigma

ing of the index when the data flows from the plugboard toward the reflector, Decode represents the encoding of an index when the data returns from the reflector toward the plugboard. Hence Encode looks at the values in the mapping and Decode uses values from the inverse of the mapping to determine the new index. The operation Substitute first encodes the index, calls Substitute on the next component in the linked list and finally decodes the result of that call and returns the answer to the caller. This nested calling of Substitute goes on until the end of the linked list is reached or a component is encountered that has different behaviour defined for this operation:

```
operations
  protected Encode: nat ==> nat
  Encode (penc) ==
    if penc in set dom config
    then return config(penc)
    else return penc;

  protected Decode: nat ==> nat
  Decode (pdec) ==
    let invcfg = inverse config in
      if pdec in set dom invcfg
      then return invcfg(pdec)
      else return pdec;

  public Substitute: nat ==> nat
  Substitute(pidx) ==
    return Decode(next.Substitute(Encode(pidx)))
  pre next <> nil

end Configuration
```

With the configurations shown in Figure 9.3, `Encode` and `Decode` behaviour is defined in Table 9.1.

Table 9.1: Example `Encode` and `Decode` behaviour

Configuration	Encode	Decode
Plugboard	Encode(1) = 1	Decode(1) = 1
	Encode(2) = 3	Decode(2) = 3
	Encode(3) = 2	Decode(3) = 2
	Encode(4) = 4	Decode(4) = 4
Rotor	Encode(1) = 2	Decode(1) = 2
	Encode(2) = 1	Decode(2) = 1
	Encode(3) = 4	Decode(3) = 4
	Encode(4) = 3	Decode(4) = 3
Reflector	Encode(1) = 2	Decode(1) = 2
	Encode(2) = 1	Decode(2) = 1
	Encode(3) = 4	Decode(3) = 4
	Encode(4) = 3	Decode(4) = 3

Exercise 9.4 Consider the following configurations:

- *Plugboard* = {1 |-> 3, 3 |-> 1}
- *Rotor* = {1 |-> 3, 2 |-> 1, 3 |-> 2, 4 |-> 4}
- *Reflector* = {1 |-> 3, 2 |-> 4}

Calculate the results for `Encode(x)` and `Decode(x)` where `x in set {1,...,4}`. □

9.4.4 Reflector

The class `Configuration` is sufficient for constructing the models for all Enigma components. The `Reflector` is an interesting case because it is at the end of the linked list, so the `next` member variable should be `nil`. Writing the invariant for the `config` member variable is more challenging. Because inputs and outputs of the reflector are on the same physical disk, each position must be an element of either the domain or the range of the mapping (the intersection of domain and range must be empty). Furthermore, the union of the domain and range must be identical to the indices of the alphabet, to ensure that all positions are accounted for. The injective property of the mapping in combination with the previous invariants now guarantees that all positions are connected in pairs. Note that no member variables have been added – only the invariant on the instance variables has been defined:

```
class Reflector
  is subclass of Configuration

instance variables
  inv ReflectorInv(next, config, alph)

functions
  ReflectorInv:
    [Component] * inmap nat to nat * Alphabet -> bool
  ReflectorInv (pnext, pconfig, palph) ==
    pnext = nil and
    dom pconfig inter rng pconfig = {} and
    dom pconfig union rng pconfig = palph.GetIndices()
```

The `Reflector` class contains three state components: the instance variables `next` (from `Component`), `config` and `alph` (from `Configuration`). Note that the last line of the invariant refers to two member variables in a single boolean expression. Care will have to be taken when updating these variables to ensure that the invariant is not invalidated. The effect of this can be seen in the `Reflector` constructor. Suppose the member variable `config` is initialised first and `alph` second. Evaluating the invariant after the first assignment will cause a run-time error because `alph` is not yet initialised. Reversing the assignment order will not solve this problem. In this sort of situation, the `atomic` statement that was introduced in Section 7.3 is used around a block statement. All statements inside the block statement are evaluated as if they are a single (atomic) statement and the invariants are evaluated only at the end of the block.

The starting position of the reflector is set only once (when Enigma is initialised) so there is no need to store that value. Instead, the mapping `pcfg` is changed such that it reflects the starting position when `config` is initialised. A map comprehension is used to modify the mapping. The variable `i` iterates over all values in the domain of the original mapping `pcfg`. This value is used to construct a new mapping where both the domain value (`i`) and the range value (`pcfg(i)`) are updated using the `Shift` operator of alphabet pa. Consider the `Reflector` configuration {1 |-> 2, 3 |-> 4} from Figure 9.3. This configuration is changed depending on the start position; the result is shown in Table 9.2.

```
operations
  public Reflector:
    nat * Alphabet * inmap nat to nat ==> Reflector
  Reflector (psp, pa, pcfg) ==
    atomic (alph := pa;
      config := {pa.Shift(i, psp-1) |->
        pa.Shift(pcfg(i), psp-1) |
```

```
        i in set dom pcfg})
pre psp in set pa.GetIndices() and
    ReflectorInv(next, pcfg, pa);
```

Table 9.2: The start position determines the configuration

Start position	Transposition distance	Configuration
1	0	{1 |-> 2, 3 |-> 4}
2	1	{2 |-> 3, 4 |-> 1}
3	2	{3 |-> 4, 1 |-> 2}
4	3	{4 |-> 1, 2 |-> 3}

Exercise 9.5 Consider the following `Reflector` configuration for a six-character alphabet: {1 |-> 4, 2 |-> 6, 3 |-> 3, 4 |-> 2, 5 |-> 5, 6 |-> 1}. Calculate the new configuration for the transposition distance 2. □

Finally, the operation `Substitute` needs to be redefined because the reflector is at the end of the linked list. The default behaviour (defined in `Component`) would cause a run-time error because `next.Substitute()` does not exist:

```
public Substitute: nat ==> nat
Substitute (pidx) ==
  if pidx in set dom config
  then Encode(pidx)
  else Decode(pidx)

end Reflector
```

9.4.5 Rotor

The cryptographic strength of Enigma is mainly due to the rotor. This device ensures that each character is encoded (information flows from plugboard to reflector) using a different transposition relation. This is achieved by turning the rotor one step before the character is substituted. Note that the rotor is not turned when the character is decoded (information flows from the reflector to the plugboard), the identical configuration is used for that. Whether the rotor turns is dependent on the position of the rotor in the linked list and the position of the latch on the rotor. The rotor nearest to the plugboard is turned for each character, independent of its latch position. The second rotor only turns if its latch is at the same position as that of the first rotor. If this is the case, the second rotor turns to its next position and then waits until the first disk has made a full rotation before turning again one step. This algorithm applies to each pair of adjacent rotors. It resembles the way that old analog odometers in cars work. Each disk has ten positions, numbered 0 to 9. If the rightmost disk moves from 9 to 0, it will also move the adjacent disk on the left to the next position and so on.

The VDM++ class `Rotor` models this behavior. Two new instance variables, `latch_pos` and `latch_lock`, are introduced to model the latch. The instance variable `latch_pos` is used to store the position of the latch on the disk. Both disks are turned one step when the latch positions of two adjacent rotors match. Note that the latches are then again in the identical position. The status is stored in the boolean variable `latch_lock` to prevent the rotor from turning instead of waiting for the adjacent disk to have made a full rotation. The rotor is idle if `latch_lock` is `true` and the latches are adjacent, otherwise it turns:

```
class Rotor
  is subclass of Configuration

instance variables
  latch_pos : nat;
  latch_lock : bool := false;

inv RotorInv(latch_pos, config, alph)

functions
  RotorInv: nat * inmap nat to nat * Alphabet -> bool
  RotorInv (platch_pos, pconfig, palph) ==
    let ainds = palph.GetIndices() in
      platch_pos in set ainds and
      dom pconfig = ainds and
      rng pconfig = ainds and
      exists x in set dom pconfig & x <> pconfig(x)
```

The invariant states that all the possible positions (represented by the indices of all characters in the alphabet) occur in both the domain and the range of `config`. Also note that "straight through" connections (e.g., input 1 is connected to output 1) are allowed, but the existential quantification is added to disallow rotors that have no effect (when all connections are "straight through" connections). At least one connection must have different input and output indices.

The constructor again must ensure that the instance variables are instantiated in a manner consistent with the invariant. It is defined as follows:

```
operations
  public Rotor:
    nat * nat * Alphabet * inmap nat to nat ==> Rotor
  Rotor (psp, plp, pa, pcfg) ==
    atomic (latch_pos := pa.Shift(plp,psp-1);
      alph := pa;
      config := {pa.Shift(i,psp-1) |->
                  pa.Shift(pcfg(i),psp-1) |
                  i in set dom pcfg})
  pre psp in set pa.GetIndices() and
      RotorInv(plp, pcfg, pa);
```

As with the Reflector, the initial position of the Rotor is only required now; there is no need to store its value. Both the latch position latch_pos and the transposition relation config are updated with the start position. Note that the latch_lock member variable is already initialised (to false) in the instance variable block.

The two overloaded Rotate operations that were defined in the base class called Component can now be refined. Note that the member variable next must be of the proper type – rotors only refer to other rotors or to the reflector. This is tested using the isofclass operator. Note that this operator checks the *type* of a variable rather than its value. The operator isofclass takes two arguments. The first argument is a class *name*, the second argument is an object reference *expression*:

```
isofclass( name, expression )
```

The operator yields the boolean value true if and only if the object reference expression refers to an instance of the same type as class name or any of the subclasses of name, and false otherwise. So, the examples

```
let a = new Alphabet("AB") in
  isofclass(Alphabet, a)

let a = new Alphabet("AB") in
let r = new Reflector(1, a, {1 |-> 2}) in
  isofclass(Component, r)
```

will yield true, whereas the examples

```
let a = new Alphabet("AB") in
  isofclass(Component, a)

isofclass(Rotor, nil)
```

will yield false.

```
public Rotate: () ==> ()
Rotate () ==
  (-- propagate the rotation to the next component
   -- and tell it where our latch position is
   next.Rotate(latch_pos);
   -- update our own latch position and take the
   -- alphabet size into account
   if latch_pos = alph.GetSize()
   then latch_pos := 1
   else latch_pos := latch_pos+1;
   -- update the transpositioning relation by
   -- shifting all indices one position
   config := {alph.Shift(i) |->
              alph.Shift(config(i)) |
              i in set dom config};
   -- remember the rotation
   latch_lock := true)
pre isofclass(Rotor,next) or
    isofclass(Reflector,next);

public Rotate: nat ==> ()
Rotate (ppos) ==
  -- compare the latch position and the lock
  if ppos = latch_pos and not latch_lock
  -- perform the actual rotation
  then Rotate()
  -- otherwise reset the lock
  else latch_lock := false
pre ppos in set alph.GetIndices();

end Rotor
```

Notice that the substitution functionality that was modelled in Configuration has not changed at all. The substitution and rotation algorithms are kept separate, sharing only the config member variable. It makes model maintenance a lot easier because the functionality is clearly separated. Table 9.3 presents a three-rotor system. Rotor 1 is connected to the plugboard and rotor 3 is connected to the reflector. The member variable latch_pos of rotor 1 is changed every time a character is encoded. Rotors 2 and 3 are turned only when they have identical latch positions and the member variable latch_lock is false. These situations are illustrated by the "⇒" symbol in Table 9.3.

Table 9.3: A three-rotor system

Rotor 1		Rotor 2		Rotor 3	
pos	lock	pos	lock	pos	lock
2	-	3	false	3	false
⇒ 3	-	⇒ 3	false	⇒ 3	false
4	-	4	true	4	true
1	-	4	false	4	true
2	-	4	false	4	true
3	-	4	false	4	true
⇒ 4	-	⇒ 4	false	4	true
1	-	1	true	4	false
2	-	1	false	4	false
3	-	1	false	4	false
4	-	1	false	4	false
⇒ 1	-	⇒ 1	false	4	false
2	-	2	true	4	false
3	-	2	false	4	false

9.4.6 Plugboard

The plugboard is used to replace pairs of characters. As with the other components, the injective mapping of indices, config, is used to represent this. The plugboard is initialised with another injective mapping that contains solely disjoint sets of domain and range elements (note the precondition of the constructor). Each maplet represents a replacement. For example, the mapping {1 |-> 2} implies that the character with index 1 is replaced by the character with index 2. To express the notion of replacement *in pairs*, the mapping is merged with its own inverse. In the example, config is {1 |-> 2, 2 |-> 1}. It is sufficient to claim that the domain of config shall be a subset of the set of indices of the alphabet, because the domain and range are identical due to the mapping union that was just performed. What happens with indices of the alphabet that do not occur in dom config? Recall that the Encode and Decode operations, which are defined in class Configuration, simply return the index if that value does not occur in either the domain or range of the mapping config. This behaviour reflects the fact that the character was not swapped by the plugboard:

```
class Plugboard
  is subclass of Configuration

instance variables
  inv PlugboardInv(config, alph)

functions
  PlugboardInv: inmap nat to nat * Alphabet -> bool
  PlugboardInv (pconfig, palph) ==
    dom pconfig subset palph.GetIndices()

operations
  public Plugboard:
    Alphabet * inmap nat to nat ==> Plugboard
  Plugboard (pa, pcfg) ==
    atomic (alph := pa;
      config := pcfg munion inverse pcfg)
  pre dom pcfg inter rng pcfg = {} and
      PlugboardInv(pcfg, pa);
```

The Substitute operation of the plugboard is modified because the first rotor in the linked list needs to be rotated before the transposition actually takes place. Also note that the member variable next needs to be of the correct type (Rotor or Reflector):

```
  public Substitute: nat ==> nat
  Substitute (pidx) ==
    (next.Rotate();
     Configuration'Substitute(pidx))
  pre pidx in set alph.GetIndices() and
      (isofclass(Rotor,next) or
       isofclass(Reflector,next))

end Plugboard
```

The operation Substitute is redefined for the third time! First it was defined in the abstract base class Component, then it was redefined in the Configuration class and now again in the class Plugboard. Although this is allowed in VDM++ it is not always easy to grasp the consequences of the redefinition. For example, not only the behaviour specifications but also pre- and postconditions are *implicitly* redefined. In Figure 9.4, the inheritance relations for the Substitute operation are shown. Note that the operation in class Reflector relaxes the precondition that was specified in the base class. In contrast, the operation in class `Plugboard` makes the precondition much stronger because it also enforces properties on the parameter `pidx`. The original precondition is again implicitly enforced because the operation `Configuration`Substitute` is called inside the body of the operation. Note that the class `Rotor`, which is not shown in the figure, has equivalent behaviour to the class `Configuration`.

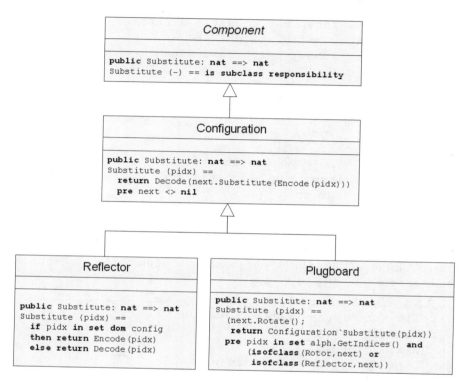

Fig. 9.4: Redefinition of the `Substitute` operation

9.4.7 Enigma

All the basic Enigma components have been modeled in the previous subsections and a simple model of the Enigma device is presented here that works on an alphabet of four characters configured as illustrated in Figure 9.3. First some values are defined that are used to initialise these components. The value `refcfg` is used for the configuration of the reflector, `rotcfg` is the configuration of the rotors and finally `pbcfg` is the configuration of the plugboard:

```
class SimpleEnigma
  is subclass of Component

values
  refcfg : inmap nat to nat =
    {1 |-> 3, 2 |-> 4};
  rotcfg : inmap nat to nat =
    {1 |-> 2, 2 |-> 4, 3 |-> 3, 4 |-> 1};
  pbcfg : inmap nat to nat =
    {2 |-> 3}
```

The linked list of components is constructed in reverse order, last element first. The Reflector is created with the new operator and the initial position; the alphabet and configuration are passed as parameters to the constructor. Note that three rotors are constructed that have identical configurations but the initial position and the position of the latch are different for each rotor:

```
operations
  public SimpleEnigma: () ==> SimpleEnigma
  SimpleEnigma () ==
    (dcl cp : Component ;
    alph := new Alphabet("ABCD");
    next := new Reflector(4,alph,refcfg);
    cp := new Rotor(3,3,alph,rotcfg);
    cp.SetNext(next);
    next := cp;
    cp := new Rotor(2,2,alph,rotcfg);
    cp.SetNext(next);
    next := cp;
    cp := new Rotor(1,1,alph,rotcfg);
    cp.SetNext(next);
    next := cp;
    cp := new Plugboard(alph,pbcfg);
    cp.SetNext(next);
    next := cp);
```

The operation Keystroke is the only visible functionality of the SimpleEnigma device. A character is inserted and the encrypted character is returned:

```
public Keystroke : char ==> char
Keystroke (pch) ==
  let pidx = alph.GetIndex(pch) in
    return alph.GetChar(next.Substitute(pidx))
pre isofclass(Plugboard,next)

end SimpleEnigma
```

The interpreter of VDMTools can now be used to observe the behaviour of the completed model. A typical interactive session could look like this:

```
Initializing specification ... done
>> create a := new SimpleEnigma()
>> print a.Keystroke('A')
'D'
>> print a.Keystroke('C')
'A'
>> print a.Keystroke('C')
'A'
>> print a.Keystroke('C')
'D'
```

The next step is to analyse the model and determine its validity. Some confidence in internal consistency has been gained using the VDMToolssyntax checker and type checker. It is now time to test the model systematically.

9.5 The VDMUnit Framework

The VDMTools interpreter, accessible through the graphical user interface, is well-suited for prototyping. For example, it can be used to explore alternative modeling strategies and to try out new ideas interactively while the model is being constructed. The syntax and type checkers provide feedback on the internal consistency of the model, while the interpreter is used to validate parts of the model, as was shown in the small example session at the end of the previous section. This way of working is very powerful because it gives the modeller immediate feedback on design decisions and forces the modeller to consider the "big picture" – the interactions between functions and operations across the model as a whole. Weaknesses are often spotted when executing the model; they can be corrected at this stage of the development process at relatively low cost.

The purpose of model interaction is to answer the informal question: Does the model work? Unfortunately, it is not possible to answer "yes" to this question in general. First, it is impossible to prove that the set of requirements that describe the proper operation of the system is complete. There is always a risk of underspecification. A famous example is a military radar system developed in the 1970s to detect incoming ballistic missiles. When the system was first turned on after an elaborate test process, alarms went off that indicated that an enemy missile was fired when nothing had happened. Missile-tracking radars had become so powerful that the moon actually reflected the radar waves. The system alarm was triggered by the moon rising above the horizon. The requirement to filter out the radar reflection from the moon had been completely overlooked by the designers. Second, even if a complete set of requirements could exist, it is impossible to prove (due to Turing's undecidability theory and the so-called halting problem) in general that a computer program will terminate for any sequence of input events. In summary, providing a proof to the consistency and completeness of a model is a *provable unsolvable* problem! Beizer [Beizer90] summarizes this as follows:

- We can never be sure that the specifications are correct.
- No verification system can verify every correct program.
- We can never be certain that a verification system is correct.

The last case proposed by Beizer hints in particular at mechanical verification (a computer program that implements a certain verification algorithm). It is impossible to prove that the tool is bug-free, even though the verification algorithm used is proven to be sound and complete. How much trust do we have in these tools? How accurate are the results they provide?

Because it is not possible to break the theoretical limit, attention moves from absolute proof to a weaker, but suitably convincing, demonstration. The overall aim is to raise confidence in the model by applying several available analysis techniques, trying to approach the theoretical limit as closely as possible. Model checking and formal proof are two such techniques but they are outside the scope of this book. Extended static analysis and external validation combining a VDM++ model with a GUI will be presented in Chapter 13. In the remainder of this section, *structured testing* will be examined.

Beizer states that the purpose of *testing* is to show that a program has bugs. In contrast, the purpose of *debugging* is to find the error or misconception in the model that led to the failure and to design and implement changes that correct the model. It is important to distinguish these terms because they are often confused. The example at the end of Section 9.4.7 used the interpreter for *debugging* rather than *testing*. Testing is a predefined procedure applied under known conditions that has a predictable outcome; debugging starts from a possibly unknown initial condition and often has an unpredictable outcome. The purpose of structured testing is to assess the quality of the model at different levels of abstraction. Beizer defines four significant levels:

Unit Testing: A *unit* is the smallest testable piece of software that can be put under the control of a so-called *test harness* or *driver*. This test harness implements the predefined testing procedure which controls the initial conditions, executes the test and verifies the outcome. Each VDM++ class can be considered such a *unit*.

Component Testing: A *component* is an aggregate of two or more units or components. Component testing aims at analysing functional and structural consistency of the aggregation of those classes and components.

Integration Testing: Several components are aggregated to create even larger components. The difference with component testing is that now consistency among components is analyzed, for example call and return sequences, data validation criteria and proper handling of data objects passed between components.

System Testing: Several components are aggregated to form the system. System testing concerns issues that can only be exposed by testing the entire system or a major part of it, such as startup and error recovery.

The general approach is to test software bottom-up, starting with unit testing, and to continue upward on the abstraction ladder until the system-level tests are reached and successfully completed. Testing at a higher level should be started if and only if all tests on the lower levels have been performed. The model should be modified if a test fails and then *all* tests should be repeated, starting again at the bottom, to guarantee that the change did not lead to inconsistencies elsewhere in the model. This process is called *regression testing*. This shows that testing is a process that should be easy to perform and repeat. The interpreter included in the graphical user interface of VDMTools is not well suited for this activity. Although it is possibile to build simple test scripts (basically a list of interpreter commands that will be executed sequentially), it is much easier to use the command-line version of VDMTools (see Section 3.4.10) in a batch-oriented style. Consider the process flow shown in Figure 9.5.

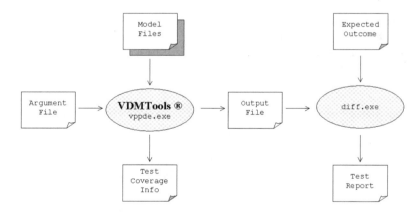

Fig. 9.5: Testing a model using the command-line version of VDMTools

The argument file contains the test case to be executed; for example, consider the file test.arg, which could read:

```
let str = "ACCC" in
  let en = new SimpleEnigma() in
    [en.Keystroke(str(i)) | i in set inds str]
```

The model files and the argument file are passed to the VDMTools command-line interpreter. Note that the interpreter is invoked with dynamic type checking (-D), invariant (-I), pre- (-P) and postcondition (-Q) checking turned on:

```
vppde -iDIPQ -O test.res test.arg Alphabet.vpp
  Component.vpp Configuration.vpp Plugboard.vpp
  Reflector.vpp Rotor.vpp SimpleEnigma.vpp
```

The file test.res will now yield the following result:

```
"DAAD"
```

The resulting output file can now be compared against a file that contains the expected outcome of the test case, for example, using the standard Unix tool "diff." The expected outcome file is normally written by hand by the tester, often together with the test case. This approach can be easily extended using other standard Unix tools such as "make" and "perl" to construct a *database* of test cases. For example, to maintain overview, a simple directory structure can be used to group test cases for a specific purpose or a specific abstraction level. "make" is used to iterate over all test cases contained in these directories and "perl" is used to create a single test report composed from all individual test results. The pragmatic approach to structured testing presented here seems very favourable, but there are some drawbacks to take into account:

1. VDMTools is restarted for each test case to ensure a clean initial state at the beginning of each test. This may lead to long waiting times when a complete regression test is performed over a large test database.
2. The comparison between the output file and the expected outcome file is only performed on the basis of textual (or binary) equality while the comparison may require some looseness. For example, the set $\{1,2,3\}$ is equivalent to $\{3,2,1\}$ because the order in the set is not relevant but it will be rejected by "diff."
3. Maintenance of the database is error prone and laborious because a test case is spread over at least two files, and it needs to be included in the management code of the database (e.g., a makefile).
4. There is no guidance for the structure of test case itself; basically it can be any arbitrary VDM++ expression. Ideally a simple and generic structure should be used for all test cases to ensure maintainability and ease of use.

The extreme programming community has recently put a lot of emphasis back on structured testing. One of the main research themes has been to improve the efficiency of the testing activity in the software life cycle. In their opinion, testing should become as easy as writing code. The problems listed here are typical examples that they want to address and solve. In this section, a solution to these problems is proposed, which is called VDMUnit. It is inspired on the JUnit framework, which was originally designed by Kent Beck and Erich Gamma. JUnit is an open-source framework for testing Java programs (see http://www.junit.org). The concepts proposed by the framework are generic and are easily transferred into other modeling and programming languages, such as VDM++. An overview of the VDMUnit framework is provided in Figure 9.6.

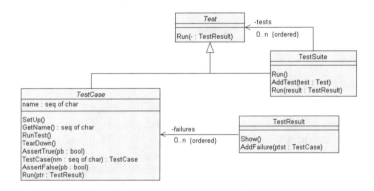

Fig. 9.6: An overview of the VDMUnit framework

9.5.1 Test

The abstract base class `Test` provides a single operation called `Run`, which is used to execute the test. The parameter of type `TestResult` is used to store the outcome of the test. `TestResult` will be presented later in this section:

```
class Test

operations
  public Run: TestResult ==> ()
  Run (-) == is subclass responsibility

end Test
```

Test is a superclass to both TestCase and TestSuite. These classes are needed to build a hierarchy of test cases, which will replace the directory structure in the pragmatic approach shown earlier. A test suite is a collection of either test cases or other test suites. Test suites can be arbitrarily nested and can consist of any number of test cases. This provides maximum flexibility as to how the test database is organised. An example hierarchy is shown in Figure 9.7.

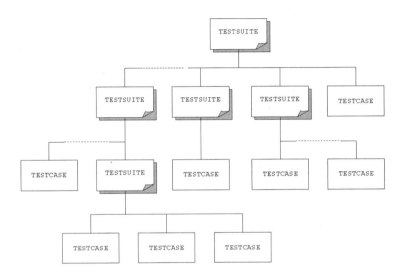

Fig. 9.7: An example hierarchy of test cases

9.5.2 TestSuite

In the TestSuite class the instance variable tests is used to store object references to all test suites and test cases contained in this test suite:

```
class TestSuite
  is subclass of Test

instance variables
  tests : seq of Test := [];
```

The first Run operation is provided for convenience; it is used only by the top-level test suite in the hierarchy. A TestResult instance is created and passed as a parameter to the second Run operation. On return, the results of the test are reported by means of a call to the operation ntr.Show:

```
operations
  public Run: () ==> ()
  Run () ==
    (dcl ntr : TestResult := new TestResult();
    Run(ntr);
    ntr.Show());
```

The second overloaded Run operation redefines the operation from the abstract base class Test. The operation iterates over all elements in the sequence tests and calls the Run operation of each element, while passing result as a parameter. The parameter result is used to "collect" the outcome of each individual test. The operation AddTest is used to add a test (which can be a test case or a test suite) to the test suite:

```
  public Run: TestResult ==> ()
  Run (result) ==
    for test in tests do
      test.Run(result);

  public AddTest: Test ==> ()
  AddTest(test) ==
    tests := tests ^ [test];

end TestSuite
```

Exercise 9.6 Redesign the VDMUnit framework so that a strict hierarchy of test suites and test cases is enforced. Currently, it is possible to create loops by adding a test suite to itself, for example. □

9.5.3 TestCase

Each test case is given a symbolic name, which makes it possible to generate a human-readable and descriptive error message if a failure is detected during testing:

```
class TestCase
  is subclass of Test

instance variables
  name : seq of char

operations
```

```
public TestCase: seq of char ==> TestCase
TestCase (nm) == name := nm;

public GetName: () ==> seq of char
GetName () == return name;
```

Two operations, `AssertTrue` and `AssertFalse`, are provided to assert test conditions. Consider the factorial function `fac` from Chapter 5, which is defined as:

```
fac: nat1 -> nat1
fac (n) ==
   if n > 1
   then n * fac(n-1)
   else 1
```

It is now possible to write the following assertions:

```
AssertTrue(fac(3) = 6)
AssertFalse(fac(3) = fac(4))
for all i in set {1,...,10} do
   AssertTrue(fac(i+1) = (i+1)*fac(i))
```

Note that the operation `assertEqual` from the JUnit framework is not supported because it is unnecessary. The equality operator "=" is defined for all types in VDM++, so it is sufficient to write `AssertTrue( `$expr_1 = expr_2$` )`, where *expr* can be an arbitrary complex expression:

```
protected AssertTrue: bool ==> ()
AssertTrue (pb) == if not pb then exit <FAILURE>;

protected AssertFalse: bool ==> ()
AssertFalse (pb) == if pb then exit <FAILURE>;
```

The assertion operations *raise an exception* if the assert condition is not met. This is done using the *exit statement*, which has the following syntax:

```
exit
```

or

```
exit expression
```

In the latter case, the expression is used to indicate what kind of exception is raised. Here, the expression <FAILURE> is used. The consequence of the exit statement is that the normal thread of control of the operation is *aborted* and execution is resumed in the innermost exception handler. The innermost exception handler is defined as the last exception handler block that was passed before the exception occurred. Note that the exception handler does not need to be located inside the same class; it can be defined practically anywhere! If the exception handler takes care of the exception, execution is resumed from the exception handler block (note: *not* the location where the exception occurred); otherwise the exception is propagated one level up and so on. A run-time error will occur if an exception is not handled at all. An exception handler block can be defined using the *trap statement*, which has the following syntax:

```
trap pattern with statement-1 in statement-2
```

First, *statement-2* is evaluated. If an exception was raised, the value of *statement-2* is matched against the *pattern*. If there is no matching, the exception is returned as the result of the complete *trap statement*, otherwise *statement-1* is evaluated and the result of this evaluation is also the result of the complete *trap statement*. Note that ptr.AddFailure is called only if an exception with the quote value <FAILURE> occurs in SetUp, RunTest or TearDown. This call will add the current test case to a list of failed test cases as will be shown later:

```
public Run: TestResult ==> ()
Run (ptr) ==
  trap <FAILURE>
    with
      ptr.AddFailure(self)
    in
      (SetUp();
       RunTest();
       TearDown());
```

The operation Run, which is redefined from the base class Test, proposes a standard approach that each test case should implement. First, the operation SetUp is called to create a suitable initial condition for the test case. Second, the operation RunTest will actually perform the test, using the AssertTrue and AssertFalse conditions. Finally, the operation TearDown will be called to clean up in such a way that the system under test can be reinitialized to perform another test case:

```
    protected SetUp: () ==> ()
    SetUp () == is subclass responsibility;

    protected RunTest: () ==> ()
    RunTest () == is subclass responsibility;

    protected TearDown: () ==> ()
    TearDown () == is subclass responsibility

end TestCase
```

Specific test cases can be constructed by inheritance from the class TestCase by redefining the operations SetUp, RunTest and TearDown. This approach solves several problems at the same time. The test case and the comparison to the expected outcome are kept together, which makes maintenance a lot easier and transparent. Furthermore, arbitrarily complex assertion statements can be written using the expressiveness of VDM++ which resolves the restrictive textual comparison of the result and expected outcome in the pragmatic test approach. Finally, a "boilerplate" solution, which is very easy to use, is provided for each test case.

9.5.4 TestResult

The class TestResult maintains a collection of references to test cases that have failed. The exception handler defined in the operation Run of class TestCase calls the operation AddResult, which will append the object reference of the test case to the tail of the sequence failures. The operation Show is used to print a list of test cases that have failed or provide a message to indicate that no failures were found. Note that the standard I/O library, which is supplied with VDMTools, is used here. IO.echo prints a string on the standard output, just like System.out.println in Java. The *def statement* is used to suppress the boolean value returned by IO.echo:

```
class TestResult

instance variables
  failures : seq of TestCase := []

operations
  public AddFailure: TestCase ==> ()
  AddFailure (ptst) == failures := failures ^ [ptst];

  public Print: seq of char ==> ()
  Print (pstr) ==
    def - = new IO().echo(pstr ^ "\n") in skip;
```

```
public Show: () ==> ()
Show () ==
  if failures = [] then
    Print ("No failures detected")
  else
    for failure in failures do
      Print (failure.GetName() ^ " failed")

end TestResult
```

9.6 Testing the Enigma Model

With the VDMUnit framework defined in the previous section it is possible to apply structured testing on the Enigma model. All the test cases that were specified can be found on the book's Web site. Only the test cases for the Plugboard and the SimpleEnigma will be shown here, as well as the top-level test suite for the model, which is called EnigmaTest.

9.6.1 PlugboardTest

Class PlugboardTest contains the test cases related to class Plugboard. A plugboard cannot be tested without either a reflector or a rotor and a reflector; therefore two operations are defined: SimpleTest and ComplexTest, respectively. The values refcfg, rotcfg and pbcfg are used to define the configurations of the reflector, rotor and plugboard:

```
class PlugboardTest
  is subclass of TestCase

values
  refcfg : inmap nat to nat =
    {1 |-> 2, 3 |-> 4};

  rotcfg : inmap nat to nat =
    {1 |-> 2, 2 |-> 1, 3 |-> 4, 4 |-> 3};

  pbcfg : inmap nat to nat =
    {1 |-> 3}
```

An alphabet is required to instantiate the reflector, the rotor and the plugboard. The instance variable `alph` is defined for that purpose, which is initialised by the operation `SetUp`. Note that this operation redefines the operation `SetUp` from the base class `TestCase` of the VDMUnit framework:

```
instance variables
  alph : Alphabet

operations
  protected SetUp: () ==> ()
  SetUp () == alph := new Alphabet("ABCD");
```

The operation `SimpleTest` constructs the smallest possible device that contains a plugboard – a plugboard directly connected to a reflector. Observe how the `SetNext` operation is used to connect the two components. The assert conditions were manually calculated and inserted here by the maintainer of the test case:

```
protected SimpleTest: () ==> ()
SimpleTest () ==
  (dcl tc : Plugboard := new Plugboard(alph,pbcfg);
   tc.SetNext(new Reflector(1,alph,refcfg));
   AssertTrue(tc.Substitute(1) = 4);
   AssertTrue(tc.Substitute(2) = 3);
   AssertTrue(tc.Substitute(3) = 2);
   AssertTrue(tc.Substitute(4) = 1));
```

The operation `ComplexTest` performs a similar task, but now with an extra component between the plugboard and the reflector, a rotor. Again, the expected outcome is calculated by hand:

```
protected ComplexTest: () ==> ()
ComplexTest () ==
  (dcl tc : Plugboard := new Plugboard(alph,pbcfg),
       rot : Rotor := new Rotor(1,1,alph,rotcfg);
   rot.SetNext(new Reflector(1,alph,refcfg));
   tc.SetNext(rot);
   AssertTrue(tc.Substitute(1) = 4);
   AssertTrue(tc.Substitute(2) = 3);
   AssertTrue(tc.Substitute(3) = 2);
   AssertTrue(tc.Substitute(4) = 1));
```

The operations `RunTest` and `TearDown` redefine the operations from the base class `TestCase`. `RunTest` calls the operations `SimpleTest` and `ComplexTest` and `TearDown` is the null operation `skip` because `SimpleTest` and `ComplexTest` only modify their own local state, due to the `dcl` declarations, which are automatically cleaned up after the operation call is completed. Note that the test case could have been split into two separate test cases; it is advisable to do this, especially if the size of the test case is larger than that shown here:

```
protected RunTest: () ==> ()
RunTest () == (SimpleTest (); ComplexTest ());

protected TearDown: () ==> ()
TearDown () == skip;

end PlugboardTest
```

9.6.2 SimpleEnigmaTest

`SimpleEnigmaTest` defines a simple yet very effective test for the Enigma model. Recall that encoding and decoding messages are reversible processes if and only if identical settings are used. Two instances of the class `SimpleEnigma` are created, and for an message of arbitrary length it is now claimed that, for all characters in the message, the decoding of a character is identical to the character itself:

```
class SimpleEnigmaTest is subclass of TestCase

operations
  protected SetUp: () ==> ()
  SetUp () == skip;

  protected RunTest: () ==> ()
  RunTest () ==
    (dcl se1 : SimpleEnigma := new SimpleEnigma (),
         se2 : SimpleEnigma := new SimpleEnigma ();
     for ch in "ABCDDCBAABCDDCBAAABBCCDD" do
       AssertTrue (
         se1.Keystroke (se2.Keystroke (ch)) = ch));

  protected TearDown: () ==> ()
  TearDown () == skip

end SimpleEnigmaTest
```

9.6.3 EnigmaTest

Class EnigmaTest is the top-level entry point of the test database. It consists of a single operation Execute, which constructs a test suite. The test suite is filled with all available test cases, and the tests are performed by calling ts.Run(). The result of the test run is eventually shown on the standard output:

```
class EnigmaTest
operations
  public Execute: () ==> ()
  Execute () ==
    (dcl ts : TestSuite := new TestSuite();
     ts.AddTest(new AlphabetTest("Alphabet"));
     ts.AddTest(new ConfigurationTest("Configuration"));
     ts.AddTest(new ReflectorTest("Reflector"));
     ts.AddTest(new RotorTest("Rotor"));
     ts.AddTest(new PlugboardTest("Plugboard"));
     ts.AddTest(new SimpleEnigmaTest("SimpleEnigma"));
     ts.Run())

end EnigmaTest
```

The test cases for the Enigma model have now been defined and the test can be performed. First, a file called all.arg is created, which contains the command to start the test run. An instance of EnigmaTest is created and the operation Execute is called:

```
new EnigmaTest().Execute()
```

The test coverage information file enigma.tc is reset before the command-line interpreter is invoked by passing all applicable specification files to the parser of VDMTools:

```
vppde -p -R enigma.tc Alphabet.vpp AlphabetTest.vpp
    Component.vpp Configuration.vpp ConfigurationTest.vpp
    EnigmaTest.vpp Plugboard.vpp PlugboardTest.vpp
    Reflector.vpp ReflectorTest.vpp Rotor.vpp IO.vpp
    RotorTest.vpp  SimpleEnigma.vpp SimpleEnigmaTest.vpp
    Test.vpp TestCase.vpp TestResult.vpp TestSuite.vpp
```

Now the command-line interpreter is invoked to start the test run. Note that the interpreter is started only *once* to execute *all* test cases. This is a major performance improvement over the pragmatic approach presented earlier:

```
vppde -iDIPQ -R enigma.tc -O all.res all.arg
   Alphabet.vpp AlphabetTest.vpp Component.vpp
   Configuration.vpp ConfigurationTest.vpp
   EnigmaTest.vpp Plugboard.vpp PlugboardTest.vpp
   Reflector.vpp ReflectorTest.vpp Rotor.vpp IO.vpp
   RotorTest.vpp SimpleEnigma.vpp SimpleEnigmaTest.vpp
   Test.vpp TestCase.vpp TestResult.vpp TestSuite.vpp
```

As expected, the following result is obtained:

```
No failures detected
(no return value)
```

Despite the fact that no failures were detected and the test suite was composed very carefully, the basic question remains: To what extent is the model really tested? A partial answer is provided by the test coverage information generated during the test run by the interpreter. Test coverage can be analysed in two ways, by generation of test coverage tables and by colouring the source text of the model. In the latter case, statements that have not been exercised by the test suite are highlighted. This makes it relatively easy to design a new test case, which is added to the test suite, that will ensure that the untouched statement is tested in the next test run. By performing a regression test and analysing the test coverage information again it is possible to verify whether the model is fully exposed to test cases. Note that it does not guarantee that the model will always work, it is just a guarantee that each statement of the model is tested by at least one or more test cases. This quality measure is an important criterion in the quest to raise confidence in the model. As an example, the test coverage information of the Alphabet class is provided in Table 9.4.

Table 9.4: Test coverage overview for the Alphabet class

Name	#Calls	Coverage
Alphabet'AlphabetInv	18	√
Alphabet'Alphabet	6	√
Alphabet'GetChar	52	√
Alphabet'GetIndex	52	√
Alphabet'GetIndices	520	√
Alphabet'GetSize	77	√
Alphabet'Shift	1366	√
Total Coverage		**100%**

9.7 Summary

This chapter has provided a detailed illustration of the construction of a model for a relatively intricate system and has introduced a systematic approach to testing as a means of gaining confidence in the model. Apart from the technical details of the Enigma device and its model, several more general points have been made, with a bearing on the use of formal object-oriented modelling techniques in practice:

- Opportunities for generalisation and reuse were exploited in making model design decisions. Patterns were identified and applied. It is worth investing time in the careful separation of generic aspects from the domain-specific aspects of the model.
- When constructing a model, it is important to bear its future maintenance in mind and let this influence structuring and abstraction decisions.
- There is a clear distinction to be made between testing and debugging. For the purposes of systematic testing, it is worth defining a test framework supporting unit tests to system-level tests and regression testing.
- Investing time in gaining maximum confidence in a model is worthwhile because of the potential gain in debugging and rework costs if major defects get through to implementation.

10

Combining Views: The CSLaM System

The aim of this chapter is to provide a deeper understanding of the relationship between UML class diagram models and the data and functional detail presented textually in VDM++. In particular, the mapping rules between models in the two notations will be clarified.

10.1 Introduction

This book has presented an approach to object-oriented system design based on the use of two system views: class structure expressed graphically in UML class diagrams and precise views of data and functionality expressed in the textual VDM++ notation. The UML class diagram notation provides a graphical overview of system structure in terms of classes and relationships between them. The VDM++ models also contain classes and relationships but augment these with descriptions of data, invariants, functions and operations expressed at a high level of abstraction. This chapter shows how UML class diagrams and a VDM++ model work together as complementary views of a system. The rules underlying the tool-supported mapping from one representation to the other are presented. The ways in which the complementary character of the models can be exploited are described and the limitations of the mapping rules are explained.

The case study for this chapter is a small system for control speed limitation and monitoring (CSLaM) for trains. As with the Enigma model, the chosen architecture reflects the physical distribution of system components. The purpose of the model is to analyse this logic of the CSLaM system with respect to the beacons along the track side and the speed of the train (see Figure 10.1).

The requirements for CSLaM will be presented gradually through the study in order to maintain a flow. However, the full collection of requirements for CSLaM is on the book's Web page. The main focus in this chapter is to present the mapping rules; these are listed in the same way as the modelling guidelines in Chapter 2.

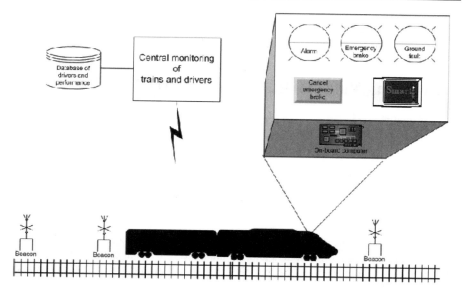

Fig. 10.1: An overview of the CSLaM system

10.2 The CSLaM System

The main purpose of the CSLaM system is to provide a continuous train speed monitoring function. This system is intended to be used when temporary speed restrictions are necessary, e.g., when repairs are taking place along a section of track. This is signalled using different types of beacons placed along the track.

The maximum permitted speed is the lower of the maximum speed of which the train is capable (a constant, e.g., 180 kilometers per hour (km/h)) and the speed limits imposed by the temporary speed restriction beacons. If the train speed exceeds the maximum permitted speed, various actions are initiated by the train's on-board computer, depending on the severity of the violation. It is ultimately possible for the on-board system to activate the emergency brake automatically if the train's excess speed is too great. Figure 10.1 shows an overview of the CSLaM system and its physical components. CSLaM is responsible for keeping track of how well train drivers adhere to the speed restrictions given by the beacons. In Figure 10.1 the box in the top right-most corner indicates the on-board parts accessible to the train driver. More specific requirements for this system are introduced later in the chapter.

In summary, the CSLaM system main components are:

1. the on-board control speed limitation (CSL) subsystem;
2. the trackside beacons; and
3. the performance monitoring subsystem.

These different components will be considered in turn.

10.3 The On-board CSL System

The on-board system for control speed limitation must ensure that the emergency brake is activated when necessary. It is safety-critical because there is a substantial risk that failure of the CSL could lead to the train functioning dangerously. The CSL system is composed of an on-board computer, a cab display, a cancel emergency braking button that can be used by the driver, and the physical emergency brake. The main classes in the CSL system are the CSL, the CabDisplay, the EmergencyBrake and the OnBoardComp classes. The relationship between these four classes is shown in Figure 10.2. For each of these classes at the UML class diagram level, there is an equivalent class at the VDM++ level.

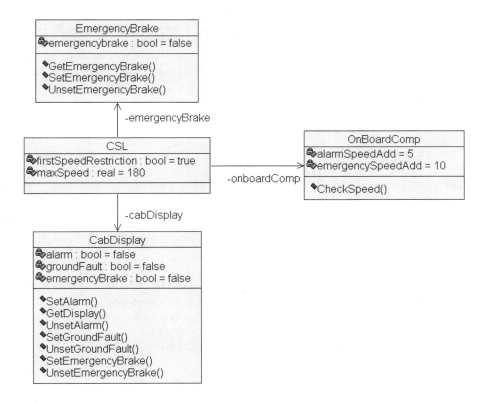

Fig. 10.2: The main classes for the CSL system

> **Mapping Rule 1:** There is a *one-to-one relationship* between classes in UML and classes in VDM++.

The three associations from the CSL class to the other classes have been provided with role names. It is a requirement for the mapping that all associations in the direction they are to be "navigated," otherwise they will be ignored by the Rose-VDM++ Link introduced in Chapter 3.

> **Mapping Rule 2:** Every *association* between two classes at the UML class diagram level must be given a *role name* in the direction(s) in which the association is to be used. This role name is then used at the VDM++ level as an instance variable whose type is given by the destination class of the association.

In the mapping VDM++ view resulting from this class diagram, all three associations are treated as *instance variables* inside the CSL class. Because none of the associations have any multiplicity the instance variables are simply declared as *object reference types* in VDM++. The VDM++ representation of the CSL class takes the following form:

```
class CSL

instance variables

cabDisplay      : CabDisplay;
emergencyBrake  : EmergencyBrake;
onboardComp     : OnBoardComp;

end CSL
```

Each instance variable provides a means of navigating the association. For example, the instance variable cabDisplay is a *state component* that is available inside the CSL class. It will always contain a reference to an instance of the class CabDisplay. Within the VDM++ view, using cabDisplay corresponds to navigating from the CSL class to the CabDisplay class via the cabDisplay association in Figure 10.2.

10.4 Member Declarations

The class diagram in Figure 10.2 includes *attributes* and the *operations* in the classes CabDisplay, EmergencyBrake and OnBoardComp. The mapping between UML class diagrams and VDM++ makes use of *stereotypes*. For attributes the value and `instance variable` stereotypes are used to indicate whether the attribute is to be treated as a constant declared in a value definition or as an instance variable.

> **Mapping Rule 3:** *Attributes* inside a UML class are represented as *instance variables* or *value* inside the corresponding VDM++ class. The stereotype will indicate the kind, and the default will be instance variables.

For UML operations the `function` and `operation` stereotypes are used to distinguish between VDM++ functions and operations. The signatures of UML operations are also taken into account in the mapping. These are used directly in VDM++ for functions and operations respectively (the only difference is the arrow used to separate the parameter types from the result type). The pre- and postconditions of VDM++ functions and operations are mapped to the corresponding elements of the UML class diagram model. However, the bodies of VDM++ operations are ignored by the mapping rules. In principle, the body could also be taken into account, but experience shows that it is easier to keep the details at the VDM++ level when working with large VDM++/UML class diagram models.

> **Mapping Rule 4:** Operations inside a UML class are represented as functions and operations in the corresponding VDM++ class, in accordance with the `function` or `operation` stereotype, and vice versa. By default the stereotype is `operation`.

At both UML class diagram and VDM++ levels each member declaration can have an *access modifier*. In Figure 10.2 these are indicated by the small *icons* in front of the member declarations. These icons are like coloured tilted boxes; `private` is indicated by a small padlock, `protected` is indicated by a small key and `public` only has the tilted box. In UML class diagrams there is also an access modifier called `implementation` that is not present in VDM++. If this is used in UML the Rose-VDM++ Link will turn it into a `private` access modifier.

> **Mapping Rule 5:** All member declarations can have *access modifiers* at both the UML class diagram; the VDM++ level and the mapping rules for these are one to one. The only exception here is that `implementation` at the UML class diagram level is considered as `private` at the VDM++ level.

For associations the access modifier is shown as a plus, a minus or a hash symbol (minus is used in Figure 10.2) to mean `public`, `private` and `protected`, respectively. Each of these modifiers is simply mapped in a one to one fashion between UML class diagrams and VDM++. The only special thing to notice here is that in VDM++ all member declarations are by default `private` whereas different defaults are used for different kinds of member declarations at the UML class diagram level. In the class diagrams in the remaining part of this chapter the member declarations will not be shown.

In VDM++ all member declarations can also be declared `static`. Inside Rose it is not possible to annotate operations to be `static`. However, all other member declarations can be annotated this way and the mapping rules simply map these parts over directly.

> **Mapping Rule 6:** Static member declarations at the VDM++ level correspond to static member declarations at the UML class diagram level, except for operations which cannot be declared static inside Rose.

All member declarations from VDM++ (except for type definitions and the concurrency constructs for synchronization and threads that do not have an equivalent at the UML class diagram level) are mapped directly using the appropriate access modifier and stereotypes.

> **Mapping Rule 7:** *Type definitions* inside a VDM++ class are simply ignored by the mapping rules because there is no equivalent at the UML class diagram level. This just means that type definitions will only be kept at the VDM++ level.

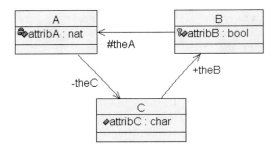

Fig. 10.3: Illustrations of different access modifiers

Exercise 10.1 Use the mapping rules for the access modifiers presented in this section to calculate the VDM++ representation manually for the classes in Figure 10.3. For this exercise and all the remaining exercises in this chapter you should then test your result *afterward* using VDMTools. □

10.5 The Cab Display

Figure 10.1 shows how the cab display is composed of three lighting indicators, an alarm indicator, an emergency braking indicator and a ground fault indicator.

The logic of the on-board system is such that it must respect rules governing the activation of the alarm indicator and the emergency brake. The requirements here are that the *alarm indicator* should be on when the train speed is higher than the alarm speed and when the two other indicators are off. The *alarm speed* is determined by the on-board computer (in this example, by adding 5 km/h to the current maximum permitted speed). The *emergency braking indicator* should be on when the train speed is higher than the emergency braking speed. The *emergency braking speed indicator* is determined by the on-board computer (in this example by adding 10 km/h to the current maximum permitted speed). The *ground fault indicator* should be on when a speed restriction beacon is not found at the distance indicated by an announcing beacon (see Section 10.6).

The attributes in the `CabDisplay` and `EmergencyBrake` classes are also modelled as instance variables in VDM++. Thus, the initial `CabDisplay` class appears as follows:

```
class CabDisplay
...
instance variables

alarm          : bool := false;
emergencyBrake: bool := false;
groundFault    : bool := false;
end CabDisplay
```

The initial value assigned to these three instance variables corresponds to the initial state requiring that all the lighting indicators are off. These initial values are taken into account in the mapping rules. This completes the treatment made of the on-board CSL system in this chapter. The entire model is present on the book's Web page.

Exercise 10.2 Use the mapping rules presented so far to calculate the UML class diagram manually for the following VDM++ classes:

```
class F

operations

public
OpInF : real ==> bool
OpInF(r) ==
  is not yet specified;

types

TinF = map bool to set of real;

end F
```

```
class G

operations

public
OpInG : () ==> bool
OpInG() ==
  is not yet specified;

end G
```

```
class H is subclass of G

instance variables
  attribH : int := 56 * valInH;

values

valInH: nat = 88;

functions

FnInH: real * real -> real
FnInH(r1,r2) ==
  r1 * r2;

end H
```

□

10.6 The Beacon Classes

Four different kinds of beacon have to be recognised by the system:

Announcement Beacon: This announces a limitation beacon. The information provided by the beacon is the target speed that must be respected when the limitation beacon is met. Several announcement beacons might be met successively.

Limitation Beacon: The speed restriction is valid as soon as the head of the train reaches the beacon. If a limitation beacon is not preceded by an announcement beacon, a ground fault is raised.

Cancel Beacon: This beacon cancels all announcements present. If no announcement is available, the cancel beacon is ignored. This beacon is used when a speed restriction is announced before a track switch and is not needed in some branches after the switch.

End Beacon: This ends the speed restriction area. The speed restriction is no longer valid as soon as the tail of the train reaches the end beacon.

The class diagram with the hierarchy for beacons is shown in Figure 10.4. All attributes and operations have been omitted.

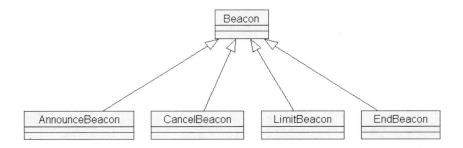

Fig. 10.4: The beacon classes

In the textual VDM++ representation, inheritance between classes is modelled using the `is subclass of` primitive. Thus, for example, the `AnnounceBeacon` class appears as follows:

```
class AnnounceBeacon is subclass of Beacon
...
instance variables

targetspeed: real;
end AnnounceBeacon
```

Mapping Rule 8: There is a *one-to-one relationship* between inheritance in UML class diagrams and inheritance in VDM++.

Exercise 10.3 Use the mapping rules for the inheritance presented in this section to calculate the VDM++ representation manually for the classes in Figure 10.5. □

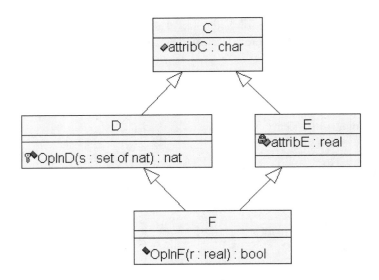

Fig. 10.5: Illustration of inheritance

10.7 Announcements and Restrictions

The on-board CSL subsystem must deal with the beacons when both the head and the tail of the train pass them. Given the length of the train, it is possible that the head of the train reaches one or more limitation beacons before the tail reaches an end beacon. All the speed restriction areas should be stored. However, there is a requirement that the CSL system cannot manage more than five valid speed restriction areas at any time. If this is violated, a ground fault must be raised.

The CSL class must keep track of the announcement beacons that have been met where the limitation beacon has not yet been met. In the same way, the CSL class must keep track of the current restrictions enforced by the currently active limitation beacons. Figure 10.6 shows how this can be represented as two associations from the CSL class to the AnnounceBeacon class and the LimitBeacon class, respectively. Note that both of them have a *multiplicity* and an ordered constraint.

At the VDM++ level, multiplicities with ordered constraints are represented using the *sequence type constructor*. This looks like:

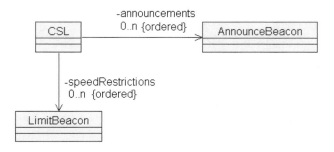

Fig. 10.6: Using ordered associations

```
class CSL
...
instance variables

announcements      : seq of AnnounceBeacon := [];
speedRestrictions: seq of LimitBeacon := [];
inv len speedRestrictions <= 5;
end CSL
```

Mapping Rule 9: The *multiplicity* of an *association* determines its type at the VDM++ level. With 0 .. n multiplicity and an ordered constraint the association is modelled as a sequence of object reference types using the *sequence type constructor*.

The multiplicity *n* is not mapped automatically and must be specified by an invariant in the VDM++. For example, an invariant is used to capture the restriction that the CSL system cannot manage more than five valid speed restriction areas at a given time.

Exercise 10.4 Use the mapping rules for the ordered associations presented in this section to calculate the VDM++ representation manually for the classes in Figure 10.7. □

10.8 Monitoring the Performance of Drivers

Having considered the basic structure of the on-board CSL system, the higher-level monitoring subsystem can be modelled. This subsystem monitors how well different

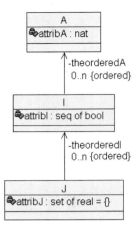

Fig. 10.7: Illustrations of ordered associations

train drivers adhere to the speed restrictions given by the beacons. To identify themselves to the system each driver has a unique smart card. On entering the train this card
must be inserted to start driving the train. Each train is also uniquely identified. These
unique identifications are used to qualify the specific driver and train, respectively.

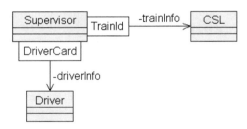

Fig. 10.8: The Supervisor class relationships

Figure 10.8 shows the top-level Supervisor class and its associations to the
Driver and CSL classes. Note how a *qualified association* is used in both cases.
The qualifiers are DriverCard (another class in the system) and the TrainId
type (defined inside the Supervisor class). In both cases this indicates a mapping
relationship from the qualified type to the class pointed to by the association. At the

VDM++ level both of these associations are treated as instance variables inside the `Supervisor` class:

```
class Supervisor
...
instance variables

driverInfo: map DriverCard to Driver := {|->};
trainInfo : map TrainId to CSL        := {|->};

types

public TrainId = token;
end Supervisor
```

Mapping Rule 10: *Qualified associations* use a type (or a class name) as the qualifier and are modelled as a *mapping* from the qualifier type to the object reference type it is going to (possibly with further multiplicity).

The automatic mapping between VDM++ and UML class diagrams has a few limitations. This, for example, prevents the user from using object reference types inside a VDM type definition introduced in Section 1.5. Thus, it would not be possible to get a correct mapping from VDM++ to UML class diagrams if, for example, the `map TrainId to CSL` type was changed to `TrainInfo`, which then was defined locally as a type inside the `Supervisor` class.

Type definitions should not be introduced for instance variables referring to object reference types. Thus, for example, one should not introduce a record type where one of the fields is a reference to the CSL class. The line one would expect on the UML class diagram would never appear if this were done. The notation for types in VDM++ is in most respects richer than that of UML class diagrams so not all VDM++ constructs have a class diagram counterpart. This is mainly a problem if complex structures arise, for example, where pairs of object reference types are used for one instance variable.

Exercise 10.5 Use the mapping rules for qualified associations presented in this section to calculate the UML class diagram representation manually for the following classes:

```
class K

instance variables

attribK   : map int to bool := {|->};
theEmap   : map int to E := {|->};
theGsmap  : map TinK to set of G := {|->};

types

TinK = seq of int;

end K
```

```
class E

instance variables

private attribE : real;

end E
```

```
class G

operations

public
OpInG : () ==> bool
OpInG() ==
  is not yet specified;

end G
```

☐

To monitor the train drivers' performance, the CSL class must take care of drivers' smart cards and keep track of faults in the current journey. Figure 10.9 shows how associations to the DriverCard class and to the Fault class are used to model this. Note that the multiplicities for these two associations are different from those seen so far.

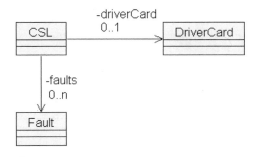

Fig. 10.9: The CSL associations for monitoring

Faults can either be sent up to the Supervisor for recording, or the on-board CSL system can record them and send them up to the Supervisor at the end of the journey. In this model the latter approach is chosen.

For the driverCard association the *multiplicity* indicates that there are *zero or one* occurrences of the DriverCard class. In VDM++ this is represented as an *optional type constructor*, as illustrated next.

For the faults association the multiplicity is indicated as *zero to n* without the ordered constraint used earlier, for example, in the announcements association. In VDM++ this is represented using a *set type constructor*. At the VDM++ level this part of the CSL looks like:

```
class CSL
...
instance variables

driverCard: [DriverCard]  := nil;
faults     : set of Fault := {};
end CSL
```

Here the VDM++ *optional type constructor* is used in the type definition for the
`driverCard` instance variable. The square brackets surrounding the object reference type `DriverCard` are the *optional type* constructor.

Mapping Rule 11: The *multiplicity* of an *association* determines its type at the
VDM++ level. 0 .. n multiplicity is modelled as an unordered collection of object
reference types using the *set type constructor*.

Mapping Rule 12: The *multiplicity* of an *association* determines its type at the
VDM++ level. Unspecified (or with 1) multiplicity is simply modelled as an object
reference type. 0 .. 1 multiplicity is modelled as an *optional type*.

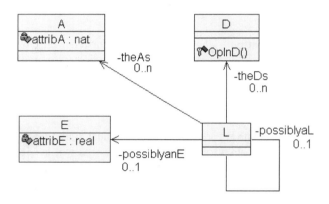

Fig. 10.10: Illustrations of unordered associations

Exercise 10.6 Use the mapping rules for unordered associations presented in this
section to calculate the VDM++ representation for the classes in Figure 10.10. □

10.9 Summary

The mapping rules between UML class diagrams and VDM++ have been demonstrated using the CSLaM system. Specific mapping rules for a number of different
concepts have been introduced:

- There is a *one-to-one relationship* between classes in UML class diagrams and
  classes in VDM++.
- *Associations* between two classes at the UML class diagram level must be given a
  role name in the direction(s) in which the association is to be used. This role name

is then used at the VDM++ level as an instance variable whose type is given by the destination class of the association.

- *Attributes* inside a UML class are represented as *instance variables* or *value* inside the corresponding VDM++ class. The stereotype will indicate the kind, and the default will be instance variables.

- *Operations* inside a UML class are represented as functions and operations in the corresponding VDM++ class, in accordance with the `function` or `operation` stereotype, and vice versa. By default the stereotype is `operation`.

- All member declarations can have *access modifiers* both at the UML class diagram and the VDM++ level and the mapping rules for these are one to one. The only exception here is that `implementation` at the UML class diagram level is considered `private` at the VDM++ level.

- Static member declarations at the VDM++ level correspond to static member declarations at the UML class diagram level, except for operations that cannot be declared static inside Rose.

- *Type definitions* inside a VDM++ class are simply ignored by the mapping rules because there is no equivalent at the UML class diagram level.

- The *multiplicity* of an *association* determines the type of which it will be declared at the VDM++ level. With `0 .. n` multiplicity and an `ordered` constraint, the association is modelled as a sequence of object reference types using the *sequence type constructor*.

- *Qualified associations* use a type (or a class name) as the qualifier and are modelled as a *mapping* from the qualifier type to the object reference type it is going to (possibly with further multiplicity).

- The *multiplicity* of an *association* determines the type of which it will be declared at the VDM++ level. `0 .. n` multiplicity is modelled as an unordered collection of object reference types using the *set type constructor*.

- The *multiplicity* of an *association* determines the type of which it will be declared at the VDM++ level. Unspecified (or with 1) multiplicity is simply modelled as an object reference type. `0 .. 1` multiplicity is modelled as an *optional type*.

Exercise 10.7 Manually calculate the VDM++ representation for the classes in Figure 10.11 using the mapping rules given in this chapter. Test your result using VDMTools. □

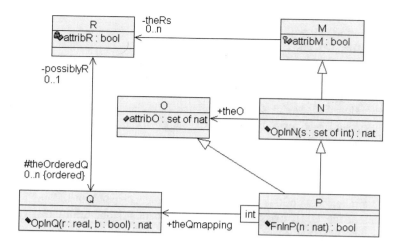

Fig. 10.11: Illustrations of UML class diagram features

11

TradeOne: From Enterprise Architecture to Business Application

This chapter aims to give an understanding of the major issues involved in applying modelling technology on the industrial scale. It does so by describing the rôle played by the VDM++ technology in the development of TradeOne, a new computer system developed by Japan Future Information Technology & Systems Co., Ltd. (JFITS). TradeOne is a *back-office solution* aiming to lower the general operating costs for trading securities. The authors are grateful to JFITS and, in particular, Shin Sahara, for their kindness and support in providing information about this application.

Perhaps the most significant issue to be addressed in developing a substantial application for a business is its architecture, in particular the relationship between the architecture of the computing system and the architecture of the enterprise into which it is to be embedded. This chapter therefore begins with a discussion of the rôle of enterprise architecture in influencing the structure of the computing system. This is then seen reflected in the models of TradeOne that JFITS developed. Finally, metrics derived from the application are presented, and the lessons learned from the application are discussed.

11.1 Introduction

The users of the TradeOne system are securities companies and brokerage houses trading in securities. A *security* is a certificate attesting credit or the ownership of stocks, options, bonds, etc. An *option* is a contract that entitles its owner to either buy or sell a security or an index at a certain price before a certain date. An *index* is a weighted average of the prices of a group of securities. The *dates* at which securities should be exercised, cancelled or agreed are critical to successful business in this domain. In existing trading companies, significant resources are therefore devoted to keeping track of the dates for securities. To remain competitive in a worldwide trade market, it is necessary to optimise the trader's business model and deal with the complexity of older traditional systems.

TradeOne is intended to lower the general operating costs and centralise the user's management resources strategically in a fast-changing infrastructure with new international trade rules. The speed with which new offers must be made in the trade of securities is increasing, for example, by shortening the securities transaction settlement date. In Japan, this has been the securities settlement date plus three business days (T+3), but there is a movement toward securities settlement date of contract plus one day (T+1). As a member of the financial global market, Japan must be able to adopt T+1 to maintain global competitiveness. This will have the effect of reducing settlement risk and drastically reducing office costs, making standard operations more efficient.

The TradeOne system automates more of the manual tasks and interfacing between different external systems, e.g., stock, trading, bank and transactions systems. In this way the business rules are automated and the organisation can focus more on conducting its trade than on administering the securities and their deadlines.

TradeOne will act as the backbone system keeping track of all transactions performed, and so it is *mission critical*: Failure could have a major impact on the user's ability to conduct business. A variety of techniques are used to control the likelihood and consequences of failures. To give increased confidence in the correctness of the TradeOne software, VDM++ was used in the development of parts of the application's kernel.

TradeOne's functionality covers several areas, shown informally in Figure 11.1. Two of these have been developed using object-oriented modelling with VDM++: the tax exemption subsystem and the options subsystem. The former automates the handling of Japanese tax regulations, until now a manual and error-prone task. The latter is responsible for handling the business process related to trading options. This kind of automated support is necessary to accommodate business process change, for example, to reduce the securities transaction settlement date.

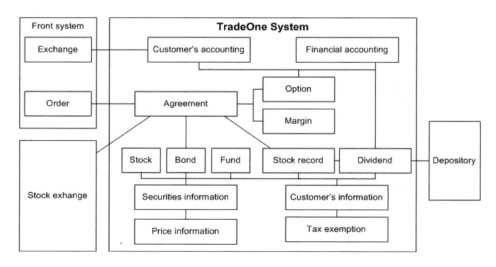

Fig. 11.1: Overview of the TradeOne system

A diagram such as Figure 11.1 can give an overview of functionality but says little about the choice of an appropriate structure for the computing system itself. For a system like TradeOne that is very closely bound to the success or failure of the business in which it is situated, it is well worth considering the relationship between the enterprise and the technical system. The following section describes the notion of an *enterprise architecture* and the process of architectural development, before the details of the design structure of TradeOne are discussed later in the chapter.

11.2 Enterprise Architectures

The success of a software development project depends crucially on the choice of a good system structure. Without a "good" structure, the software system will not withstand maintenance, changing requirements, new developments, integration with other systems, Web-enabling and many other changes during its life. Given the importance of structure, it is worth asking what the rôle of architecture is in the development of software systems and how the enhanced object-oriented design approaches presented using VDM++ fit in.

11.2.1 What Is Architecture?

There are many definitions of architecture in the computing and software context. Shaw and Garlan [Shaw&96] cite the following definition, developed at the Software Engineering Institute (SEI) in 1994:

> The structure of the components of a program/system, their interrelationships, and principles and guidelines governing their design and evolution over time.

The definition as laid out in IEEE 1471-2000 (Recommended Practice for Architectural Description for Software-Intensive Systems) [IEEE1471-2000] is:

> ... the highest level concept of a system in its environment. The architecture of a software system (at a given point in time) is its organization or structure of significant components interacting through interfaces, those components being composed of successively smaller components and interfaces.

These definitions interpret software architectures as the structure and organisation of computer programs. However, the term can usefully be interpreted in a broader context. Software is never developed in isolation: It serves a purpose, affects people and uses resources. These influences, or their consequences, have therefore to be taken into account in some way. To this end, a distinction is made between *enterprise architecture* and *software architecture*. Enterprise architecture addresses the overall architecture used by an organisation in support of some aims; the activities carried out for achieving those aims are referred to as the organisation's *business process*. An enterprise architecture includes business aspects, supporting software systems and supporting IT infrastructure. Software architecture addresses the architecture of a single software system or software system component. The impact of choosing a particular enterprise architecture is different from that of choosing particular software architecture. The ways in which each is constructed are different too. The impact of an enterprise architecture is potentially enormous on the business that an organisation does. For example, a poorly chosen or poorly implemented enterprise architecture can prevent an organisation from growing to meet increased market demands. As long as the software system is not considered company critical, the impact of a poor software architecture may be contained, albeit at some cost.

Enterprise Architectures

Enterprise architectures can be further subdivided into *business architectures*, which describe what is expected from the business process, which model is used for carrying it out and the elements needed to implement it; *software systems architectures*, which describe the software systems, and the interactions between them, providing the functional services needed to support the business process and *technical infrastructure architectures*, which describe the technical architecture providing the services, including "soft" infrastructure such as middleware, needed to run the software systems. The services to be provided by the software systems in the software systems architecture are determined by requirements that follow from the business architecture, and a similar relation holds between the software systems architecture and the technical infrastructure architecture.

The design of a single software system is ultimately determined by the business architecture part of the enterprise architecture. In this book the emphasis is on software systems. Nevertheless, the importance of enterprise architecture, the raison d'être of the software system being developed, is always kept in mind.

11.2.2 Developing Enterprise Architectures

Several techniques, such as the Zachmann framework [Zachmann87], have emerged for the development of enterprise architectures. The approach outlined here originates with Cap Gemini Ernst & Young. It is based on the premise that the construction of an enterprise architecture should be influenced by: the stakeholders affected by the system under development; the business principles that steer the company's behaviour (e.g., a policy of competing on price will lead to limits on IT budget) and its IT principles (e.g., a requirement for servers to use a particular operating system).

Exercise 11.1 There are usually more stakeholders than one might think. Consider, for example, a policy developed by the national or regional Department of Transportation, aimed at reducing road congestion. Think of at least five stakeholders. □

For each architecture (business, software systems and technical infrastructure), development follows a three-phase model:

- the *conceptual* **phase** is focused on *what* will be delivered by the architecture.
- the *logical* **phase** is focused on *how* the identified functionality will be delivered by the architecture.

- **the *physical* phase** considers *what* technology and products will be used to deliver the components of the architecture.

In each phase the identified business and IT principles can influence the architecture. Competing principles may have to be reconciled. For example, satisfaction of one principle may require a low-cost solution whereas another demands high-quality fast components. The result of the development consists of a number of schemas, each describing a stage of development of one of the architectures.

11.2.3 An Example: A Four-Tier Enterprise Architecture

Suppose a small company selling sporting goods has salespersons who travel a lot and usually work from home. Most staff have high-speed Internet connections at home, paid for by the company, allowing them to keep up with the rapid developments in the sports world using the Internet. In addition, they have just opened a branch office in another part of the country. A virtual private network (VPN) over a conventional, analogue phone line connects the local area network (LAN) in the main office to that in the back office. The company has an advanced sales system, especially developed for them using Java technology. Although the system works well and certainly provides them with competitive advantages, they regret the high maintenance costs and would have preferred a commercial off-the-shelf (COTS) system.

The company now requires that salespersons in the branch office and at home be able to use this system. The business and IT principles they have established are:

- Flexibility and readiness for the future are very important.
- The company wants to use COTS software as much as possible.
- Limit the number of different technologies in use.
- One-off costs are, depending on the fluctuating market for sportswear, always negotiable, but recurring costs such as leased line rental are not.

The need for flexibility is a powerful driver in the choice of architecture. It means that parts of the software can easily be added or replaced with minimal effects on other parts of the software. A solution with components that are responsible for a well-defined part of the functionality and limited but well-defined interfaces to other components (the "high cohesion, loose coupling" paradigm) is a good way to go about this. Furthermore, it is consistent with the use of COTS software, which also often focuses on well-defined limited functionality.

These principles, and the fact that the system required is distributed, lead to a decision that the architecture should be based on four "tiers" or layers:

- **a presentation layer,** containing the functionality to handle input from, and send output to, the end user.
- **a presentation integration layer,** containing functionality to move information between the presentation and business logic layers.
- **a business logic layer,** containing functionality relevant to the system, based on the end user's input.
- **a data layer,** containing the functionality to store and retrieve persistent data.

This four-tiered architecture, illustrated in Figure 11.2, is an example of an architectural *style*, used for many other applications in many other domains.

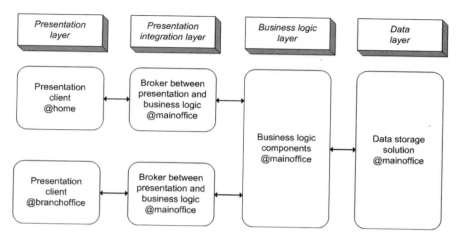

Fig. 11.2: The four-tiered architecture

At the logical level, the architecture is guided by the principle that the number of different technologies in use should be as limited as possible. This implies that, where new software should be developed, Java technology should be used and no new recurring investments can be made. On the home access side, the use of Java technology is the basis for the decision to make this possible using a simple Web browser from home, implementing the presentation integration layer using Java servlets and Java server pages (JSPs). On the branch office access side, there is the problem that the telephone connection between the branch office and the main office cannot be replaced by a leased line because of the recurring cost. Hence, a solution needs to be sought that minimises the amount of data being transported between the branch and the main office. Finally, for the data layer, a simple solution is put in place: a database. The logical level of the four-tiered architecture is represented in Figure 11.3.

At the physical level, the components from the logical architecture are replaced by real products, following the wish to use as much COTS software as possible. For example, Internet Explorer might be chosen for the home Web browser, with the iPlanet Web server for the access from home. For branch office access, Citrix MetaFrame might be chosen as it can imitate client-server applications in such a way that data transfer between the client and the server is minimised. In the main office, BEA WebLogic could be deployed as an application server. Because this product is also a servlet engine, this eliminates the need for a separate product in the presentation integration layer. Finally, Oracle 8i might be chosen as a robust and reliable database. The physical level of the four-tiered architecture is represented in Figure 11.4.

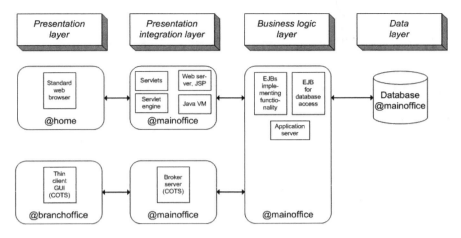

Fig. 11.3: Logical level of the four-tiered architecture

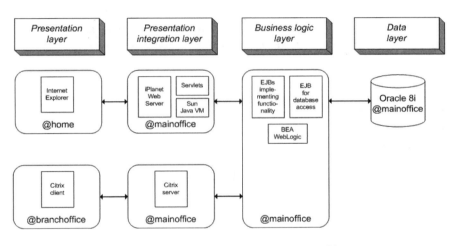

Fig. 11.4: Physical level of the four-tiered architecture

11.3 The TradeOne Architecture

The architecture of the TradeOne system is shown in Figure 11.5. This follows the four-tiered structure described in Section 11.2.3. Modelling in VDM++ was applied to the business logic and database layers, but not to the presentation layer or the presentation integration layer. This approach, in which modelling effort is concentrated on a kernel part of the application for which the highest integrity levels are required, is common.

It is important that the architecture is flexible so that it is easy to extend it with new business processes and databases. With this in mind, TradeOne is specified as a finite

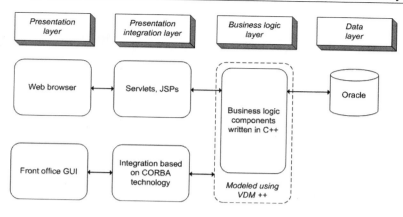

Fig. 11.5: TradeOne overall architecture

state machine, in which databases are declared as attributes and business logics (i.e., the services the system provides) are defined as operations. Thus, at a high level of abstraction, one can think of TradeOne as having the following structure:

```
class TradeOne
...
instance variables
protected db1 : DataBase1;
...
protected dbN : DataBaseN;
operations
public BusinessLogic1: ... ==> ()
...
public BusinessLogicM: ... ==> ()

end TradeOne
```

Modelling TradeOne involves developing models of `DataBase` and `Business-Logic`, as well as `Designation`, which is a collection of predicates, functions and operations used in the models of `DataBase` and `BusinessLogic`. The part of TradeOne described in this chapter deals only with two different kinds of databases: *positions* a customer can have about interest in a security (option trade) and *exercises*, which are options on which a customer has decided to exercise rights (from a position record) to purchase a particular security. The real system has more databases, but this simplified model should be sufficient to understand the principles adopted in the full model.

The model has a layered structure, with the lowest levels providing basic record structure definitions, the intermediate levels adding database access methods and business logic and the topmost level describing test cases. Disregard the top layer for the time being; it is discussed in Section 11.4. The class diagram for the positions and exercises databases, excluding the test cases layer, is shown in Figure 11.6.

The "RecordDefinition" classes provide the basic record structure for the position and exercise databases. The PositionDBBasic and ExerciseDBBasic classes provide the basic functionality for database access. All the basic databases also contain three instance variables containing the original records, the records that have been deleted and the records that have been appended. These three are needed to allow rewinding of transactions that are cancelled after they have been performed. The PositionDBPractical and ExerciseDBPractical classes provide practical interfaces to their respective databases with domain-specific functionality. The PositionEnvironment and the PositionDesignation classes provide the functionality to open, close and change positions and exercise rights to purchase securities. Thus, they form a high-level interface to the databases. Finally, the BusinessLogicClass class provides the rules that govern trading in this business area. The full system is a combination of several such databases and business logics following the same general pattern: A class for record definitions is at the top with a subclass giving a basic interface, and a further subclass giving a practically useful interface.

11.3.1 The Database Architecture in Detail

The database architecture follows the same structure as that outlined in Figure 11.6. Consider the functional models in a little more detail.

Basic Record Structure

In this class, three types, Key, Attribute and Record, and one function called KeyMatch for matching a given key and record are declared. Alternatively the Key elements could be considered domain elements in a mapping, assuming that they are unique. This would probably yield a more abstract model where the KeyMatch function could be omitted entirely:

```
class RecordDefinition
types

public Key :: ...;
public Attribute :: ...;
public Record ::
        key  : Key
        attr : AttrPart;

functions
```

```
public KeyMatch: Key * Record -> bool
KeyMatch(key,rec) == ...;

end RecordDefinition
```

For business applications, the `AttrPart` will almost always include references to one or more dates. The notions of dates and numbers of days between dates are central to the core calculations that the option subsystem performs. It is appropriate to show a small extract from the `Date` class. In the real trading system, we track the market(s) in which the trade is taking place and the appropriate weekends and holidays. This is an explicit definition of a function calculating how many days there are between two dates:

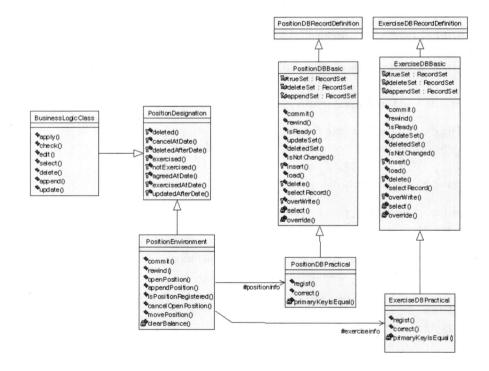

Fig. 11.6: Class diagram from TradeOne illustrating layers

```
class Date
...
types
public dayOfWeekName = <Mon> | <Tue> | ... ;

values
public lengthOfWeek = 7;

functions
public
HowManyDayOfWeekBetween2Date: Date * Date * dayOfWeekName
                                -> int
HowManyDayOfWeekBetween2Date(date1, date2, dayName) ==
  let dayNumber = DayNumberFromDayName(dayName),
      startDate = min(date1,date2),
      endDate   = max(date1,date2),
      dayCount  = DayCountBetween(startDate,endDate) + 1,
      quotient  = dayCount div lengthOfWeek,
      remainder = dayCount mod lengthOfWeek,
      dayStart  = dayNumber(startDate),
      delta     = if SubtractDayNumber(dayNumber,
                                       dayStart) = 1
                  then 1
                  else 0
  in
    remainder + delta
post
  let startDate = min(date1,date2),
      endDate   = max(date1,date2)
  in
    exists designatedDateSet : set of Date &
      forall aDate in set designatedDateSet &
        startDate.LessEqual(aDate) and
        aDate.LessEqual(endDate) and
        dayName = GetDayName(aDate) and
        RESULT = card designatedDateSet;

SubtractDayNumber: int * int -> int
SubtractDayNumber(x,y) ==
  if x >= y then x - y else x - y + lengthOfWeek;

end Date
```

Note that the postcondition for the function HowManyDayOfWeekBetween2Date uses a type binding in the exists expression. This cannot be executed in the VDM++ interpreter because it involves a potentially nonterminating search, so here it would not be possible to switch on postcondition checking while interpreting an invocation of this operation.

Basic Database Structure

Basic database functionality is provided by the class DataBaseBasic, a subclass of RecordDefinition:

```
class DataBaseBasic is subclass of RecordDefinition
...
instance variables

protected trueSet   : RecordSet := {};
protected deleteSet : RecordSet := {};
protected appendSet : RecordSet := {};
inv forall rec1,rec2 in set trueSet &
      rec1.key = rec2.key => rec1 = rec2;
```

In the TradeOne model, this level of class definition includes the declaration of basic operations over DataBase including selection, insertion, update and deletion of records:

```
operations

public Select: Key ==> RecordSet
Select(key) == ...;

public Insert: Record ==> ()
Insert(rec) == ...;

public Delete: Key ==> ()
Delete(key) == ...;

public Update: Record ==> ()
Update(rec) == ...;


...
end DataBaseBasic
```

Declaration of Practical Database Interfaces

The next level of class definition, called the "practical interface" class, defines operations over those of `DataBaseBasic` and intended for use by business logics. The tax exemption database, for example, contains an operation to insert a new record into the database described as follows:

```
class TaxExemptionDBPractical is subclass of
      TaxExemptionDBBasic
...
operations

public RegisterApplyAmt: Key * Money ==> ()
RegisterApplyAmt(key,aplAmt) ==
  def newRecord = mkRecord(key,aplAmt,normal)
  in
     Insert(newRecord)
pre true
post ApplyAmtRegistered(key,aplAmt);

...
end TaxExemptionDBPractical
```

Note how the operation, presented at the level of the business logic, uses the lower-level insertion operation defined in the corresponding basic database class. The practical interface classes then make it possible for the business logic classes to present their functionality at a uniform and appropriate level of abstraction.

11.3.2 The Business Logic Architecture

The business logic classes have no persistent data. Instead they interact with several databases and other business logics to carry out their "business". The model of each business logic is made by the combination of *scenarios* and *transitions*. The scenarios are defined implicitly using postconditions, whereas the transitions are defined explicitly. A scenario describes the scene where `BusinessLogic` is applied. A scenario consists of the following parts:

- `BusinessLogic` to be applied;
- the exception to be raised in `BusinessLogic`;
- the premise that holds before applying `BusinessLogic`;
- the conclusion derived from applying `BusinessLogic` at the scene where the premise holds.

In the VDM++ description of a scenario, an operation *observes* the application of some business logic by the `Apply` operation. A precondition will give the premise and a postcondition can be used to check the conclusion. The exception handling is accomplished by special VDM++ statements placed around the call to `Apply`:

```
class Scenario
...
operations
public Observe: BusinessLogic * ... ==> bool
Observe(BL,...) ==
  always return false
  in trap ExpectedException with return true
     in (BL.Apply(...);
          return false
        )
pre Premise
post Conclusion;

end Scenario
```

The two exception handling statements `always` and `trap` used in the body of `Observe` should be seen in connection to a statement called `exit`, which is used to raise an exception. Such exceptions can be raised inside the `Apply` operation. If an anticipated exception has been reached during the execution of `Apply`, it will be caught by the *trap statement* and `true` will be returned. If an unanticipated exception is raised inside the body of the `Apply` operation, the `always` statement will catch it and cause `false` to be returned from `Observe`. Thus, the observation should always yield `true` if the `Apply` operation is behaving as expected. Note that exit and trap statements were introduced in more detail in Section 9.5.3.

Consider the definition of the transition for a business logic:

```
class RegisterTaxExemptionApplyAmt is subclass of
       TaxExemptionDesignation

operations

public Apply: TaxExemptionDBPractical * Key * Money ==>
              ()
Apply(DB,key,aplAmt) ==
  if ProperTaxExemptionApplyAmt(aplAmt)
  then def recSet = DB.Select(key)
       in cases card recSet:
            0 -> DB.RegisterApplyAmt(key,aplAmt),
            1 -> let oldRec in set recSet
                 in if AbolishedClient(oldRec)
```

```
                         then DB.ReRegisterApplyAmt(key,
                                                   aplAmt)
                         else exit <ClientNotAbolished>,
              others -> exit <TaxExemptionKeyDuplicated>
           end
    else exit <ImproperTaxExemptionApplyAmt>;
end RegisterTaxExemptionApplyAmt_1
```

Note how different exceptional cases are treated in the body of this `Apply` operation using the `exit` statement. Scenarios can then be defined to ensure that the different logical parts of the operation have been exercised at least once.

A `Scenario` for this transition is as follows:

```
class RegisterTaxExemptionApplyAmt_1 is subclass of
      TaxExemptionDesignation

operation

public Observe: RegisterTaxExemptionApplyAmt *
                TaxExemptionDBPractical *
                Key * Money ==> bool
Observe(BL,DB,key,aplAmt) ==
  always return false
  in (BL.Apply(DB,key,aplAmt);
      return true
     )
pre DB.Ready() and
    ProperTaxExemptionApplyAmt(aplAmt) and
    def recSet = DB.Select(key)
    in
       recSet = {}
post RESULT = true and
     DB.ApplyAmtRegistered(key,aplAmt);
...
end RegisterTaxExemptionApplyAmt_1
```

Note how the precondition is used to state the premise of the scenario and how the postcondition is used to state the conclusion. Other `Scenarios` taking the different exceptional cases into account can be defined using the same principles.

11.4 The Validation Approach

A level in the TradeOne architecture handles testing. The general structure of such test cases is as follows:

```
class TestCase
...
operations
public Run: Scenario * BusinessLogic * ... ==> ()
Run(SN,BL,...) ==
  let DataDeclaration
  in
    def result = SN.Observe(BL,...)
    in
      Closing;
...
end TestCase
```

The `TestCase` class has a `Run` operation to start validation, where the data necessary for the validation and the `Scenario` and the `BusinessLogic` to be observed must be present. This could, for example, look like:

```
class RegisterTaxExemptionApplyAmt_1_1
operations

public
Run: RegisterTaxExemptionApplyAmt_1 *
     RegisterTaxExemptionApplyAmt *
     TaxExemptionDBPractical ==> ()
Run(SN,BL,DB) ==
  (SetUp();
   let key = (new TaxExemptionTestKeyDefinition()).
             NewCommer,
       aplAmt = 3000000 in
     def rslt = SN.Observe(BL,DB,key,aplAmt)
     in CloseDown(rslt));

end RegisterTaxExemptionApplyAmt_1_1
```

11.5 Metrics from TradeOne

To provide insight into the effect of modelling in the development of the TradeOne subsystems, various metrics were recorded. Size metrics for the entire system and details for the two subsystems modelled in VDM++ are presented. Information on the timing of the development and the background of the development teams is given. Finally, defect metrics for the development of the two subsystems are presented. In the development of both subsystems the final implementation was hand-coded by programmers. It is clear that higher productivity could be gained by automatically generating code. However, this was abandoned because the generated code conflicted with the RogueWave libraries used as the interface layer to Oracle.

To permit comparisons between TradeOne and other projects, standard measuring principles based on the constructive cost model (COCOMO) [Boehm81] are followed. In COCOMO the size of an application can be measured in "delivered source instructions" (DSI). The term "delivered" means that support software, such as test drivers, that is not delivered as a part of the real application is excluded unless it is developed with the same care as the delivered software, with its own reviews, test plans, documentation, etc. The term "source instructions" includes all program instructions created by project personnel and processed into machine code by some combination of preprocessors, compilers and assemblers. This excludes comments and unmodified utility software. "Instructions" are defined as lines of code.

11.5.1 Overall Size of TradeOne

DSIs are presented for the overall TradeOne system and the parts modelled using VDM++ in Table 11.1. The entire TradeOne system is substantial, and it is clear that only a small portion of the total number of DSIs result from the two subsystems modelled using VDM++. A deeper analysis of the sizes of the VDM++ models and the corresponding C++ implementations follows.

Table 11.1: Size of TradeOne

System	DSI (C++)
Total TradeOne	1,342,858
Tax exemption subsystem	18,431
Option subsystem	60,206

The Tax Exemption Subsystem

The size of the overall VDM++ model for the *tax exemption* subsystem and the classes used for systematically testing it, are 11,757 DSIs. This is broken down in Table 11.2. The DSI counts for the C++ and Java code derived manually from it are presented in Table 11.3. The C++ DSI count does not include the test bed or testing utilities. The Java code developed was a generator for screen input/output.

Table 11.2: Breakdown of tax exemption subsystem model

DSI (VDM++)	Model of
8,102	business logics
1,539	constraints
1,342	utilities for TradeOne system
774	utilities for general purpose
11,757	in total

Table 11.3: Tax exemption subsystem: Source lines per programing language

DSI	Language of
9,028	C++
9,403	Java
18,431	in total

The Option Subsystem

The size of the overall VDM++ model for the *options* subsystem and the classes used for systematically testing it is 68,170 DSIs. This is split into different categories as shown in Table 11.4. A similar breakdown of the C++ code manually derived from the model is shown in Table 11.5. The C++ DSI counts presented in the table do not include:

- any of the test scenarios.
- the test case program generated by test data generator. Only the "input data for test cases" to the test data generator are included. Strictly speaking, the test data generator does not generate test cases; it changes them into test cases (C++ programs from the Excel test data inputs.)
- utilities and their test cases. A number of C++ libraries used for TradeOne have already been developed by preceding projects. These are consequently not included either.

A proper comparison between the sizes of the VDM++ and the corresponding C++ implementation can only be made for the business logic and the business logic libraries. This also includes the database interfaces at the C++ level, because virtual database interfaces in VDM++ were included in business logic libraries.

For the options subsystem there are 18,501 VDM++ lines in total for the business logics and at the C++ level it is 48,029 lines. A plain comparison here shows that the VDM++ is 38% of the size of the C++ implementation. The main VDM expert at JFITS feels that, given the time, it would have been possible to reduce the VDM++ model to about half the stated size, mainly because of the extent to which some functionality is repeated in the model, rather than being centralised in common functions or operations.

Table 11.4: Breakdown of options subsystem model

DSI (VDM++)	Model of
10,846	business logics
7,655	business logic libraries for option system
13,771	test scenarios
31,641	test cases
153	test bed
689	test data
872	utilities for TradeOne system
627	test cases for utilities
956	utilities for general purpose
960	test cases for the utilities
68,170	in total

Table 11.5: Breakdown of options subsystem implementation (C++)

DSI	Implementation of
27,557	business logics
9,761	business logic libraries for options system
8,437	test cases for `TestCase` converter to C++
4,730	a framework called GOFO (good-old fashioned office)
3,201	a framework called Phantom of JSP for reporting
10,691	DB interfaces
64,377	in total

11.5.2 The Development Team and Project Time Scales

It is important to bear in mind the experience and skill levels of developers when considering the effects of applying a particular technology. The project time scales also give a sense of the environment in which the approach was implemented.

The Tax Exemption Subsystem

The tax exemption subsystem was developed by a small team of developers during 2000. There were six persons on this development team, averaging four persons per month for three and a half months. The development started with an initial design phase in which a system architect derived a first overall framework for the VDM++ models. After this, four VDM++ experts developed models in parallel. All the VDM++ models were reviewed by all the members of the project team. Finally two expert programmers implemented the system manually from the VDM++ model.

All the developers were more than 40 years old and each had been developing software for more than 20 years. However, for all of them it was their first development using C++ and Java. For all the developers, with one exception, it was also the first application in the financial domain. Four of them had some basic prior knowledge of formal methods and so had little difficulty applying VDM++ on this subsystem.

The Options Subsystem

The options subsystem was developed by a larger team, with 10 developers during 2001. The development started with a group of domain experts writing the functional requirements for the subsystem in Japanese. In parallel, one of the VDM experts from the tax exemption subsystem taught three new programmers in VDM++ (in less than one week). Then the domain experts wrote UML use cases with support from an expert in object orientation. In parallel with this, a system architect (the one who had worked on the tax exemption subsystem) wrote the overall framework of the VDM++ model for this subsystem. Following the completion of the use cases and the VDM++ framework, two VDM++ experts and the three newly trained programmers completed the VDM++ model concurrently. Then the model was validated using the VDM++ interpreter, and its kernel was reviewed. Finally four programmers wrote C++ and Java from the VDM++ model. Because of time pressure it was necessary to use one of the domain experts to implement in C++, and the VDM experts had to write test cases in C++.

Five of the developers also worked on the tax exemption subsystem. Two were domain experts with a Cobol background. Two were new employees, and one person was another development engineer with a Cobol background. The average age of team members was around 40. Four of the new developers had C++ experience, but they had no experience writing large C++ systems themselves.

It is worth remarking that a Japanese version of VDMTools supporting Japanese identifiers was used in the options subsystem development. This was considered important because a larger group of people was involved in the VDM++ modelling. On average, 9.5 person-months was spent every calendar month. The actual development took around seven months. The division of the effort into phases is as shown in Table 11.6.

Table 11.6: Effort profile for options subsystem

Person-months	Phase
2.8	Education
25.0	Analysis
7.0	Design
16.4	Implementation (including unit test)
8.9	Test
60.1	In total

The profile of the development shows a greater proportion of the development effort being spent in the early analysis phase than is traditionally expected. This is typical for applications of modelling technology where the quality of thorough analysis up front is rewarded in the later phases.

11.5.3 Defect Metrics

Defects discovered during the development of the TradeOne system have been recorded and categorised. Each of the two subsystems is considered in turn.

The Tax Exemption Subsystem

No defects were discovered in the system test of the tax exemption subsystem and, at the time of writing, no defects have been discovered in the running software. The defects discovered during integration testing were classified by the JFITS engineers as shown in Table 11.7

Table 11.7: Categories of defects discovered for the tax exemption subsystem

Cause of defect	Number of cases
Misunderstandings with other subsystems	9
End-user changing requirements	1
Defect in informal Japanese specification	1
Misunderstanding of terminology	1
Total	**12**

The Option Subsystem

The VDM++ model of the option system was regression tested, and test coverage was measured by the VDM++ interpreter and the handmade C++ tool plus rational code coverage. Branch coverage is more than 95% and the basic VDM++ model has 100% statement coverage. However, reviewing was only carried out on a few parts of the option system due to lack of time.

The defects discovered in integration testing were categorised as shown in Table 11.8.

Table 11.8: Categories of defects discovered for the options subsystem

Cause of defect	Number of cases
Misunderstandings with other subsystems	2
Misunderstandings of existing subsystems	4
Defect in informal Japanese specification	6
Defect in VDM++ model	3
Missing specification document of font	1
Defect in coding phase	26
Defect of making test data	1
Total	**43**

Comparison against Size of Implementation

In the option subsystem, defects occurring in the analysis and design phases were lower than for the tax exemption subsystem. A comparison between the subsystems indicates that 25% (15/12) more defects were found in the analysis and design phases. Note that for the tax exemption system, all defects were found at this level. However, the comparison at DSI level shows that the size of the option subsystem is 327% (60206/18431) greater than that of the exemption subsystem. The other interesting figures here are the defect rates compared to the size of each subsystem. Here the figure for the tax exemption subsystem shows that 0.65 (12/18431) defect is present per 1000 DSI (KDSI). For the options subsystem 0.71 (43/60206) defect is present per KDSI. Both of these are very low defect rates. The overall average for the TradeOne system is 1.12 defect per 1000 DSI. To compare to defect rates elsewhere, the order of defects in NASA software is reputed to be around around 0.1 defect per 1000 DSI, and at least 10 times normal development costs are required to reach this level of correctness. For very highly tested code the figure is around one defect per 1000 DSI. For high-quality development released code has on the order of three defects per 1000 DSI. For normal-quality commercial released code, the figure is around 30 defects per 1000 DSI. This comparison suggests that the correctness reached in the two subsystems developed using VDM++ is impressive.

11.6 Lessons Learned from the TradeOne Project

Using the COCOMO principles it is possible to create an estimate of how much effort the development of a software subsystem would require. The estimate is calculated based on the size of the application, the mode in which it is to be used and a number of additional parameters taking, for example, experience of developers into account. This calculation has been carried out for both subsystems presented here using a COCOMO calculator produced by HyperCard. For the tax exemption subsystem the calculation parameters and the corresponding results are presented in Table 11.9.

The development of the tax exemption subsystem took 14 person-months, about 36% (14/38.5) of the estimated effort based on the COCOMO model. In addition, the development took three and a half months compared to the estimated nine months from the COCOMO model (39% of the estimated duration). The development of the tax exemption subsystem appears to have been markedly better than the COCOMO estimates would suggest.

For the options subsystem the calculation parameters and the corresponding results are shown in Table 11.10

The development of the options subsystem took 60.1 person-months, 40% of the estimated required effort based on the COCOMO model. In addition, the real development took seven months, compared to the estimated 14.3 months from the COCOMO model (49% of the estimated duration). It is thus possible to conclude that development of the options subsystem is also considerably better than the estimates from the COCOMO approach.

Table 11.9: COCOMO calculation drivers for tax exemption subsystem

Area	Parameter/result
DSI	18431
Mode	semidetached
Cost Drivers	
Required software reliability	normal
Database size	normal
Product complexity	normal
Execution time constraint	normal
Virtual machine volatility	normal
Computer turnaround time	low
Analyst capability	very high
Applications experience	low
Programmer capability	very high
Virtual machine experience	normal
Language experience	low
Modern programming practices	high
Use of software tools	high
Required development schedule	very low
Result	
Effort required	38.5 person-months
Development schedule	9 months

11.7 Summary

In this chapter a real-world project using the VDM++ technology has been presented. An overview of the application and its architecture has been given. A few extracts from the VDM++ model have been used to indicate the style used and the validation principles applied. Finally, a few metrics from the use of VDM++ in this project have been shown and lessons learned have been gathered. Compared to the figures derived using the COCOMO model the productivity metrics are very impressive. In addition, the metrics for defect rates are very encouraging. Based on the results from these two subsystems, JFITS wish to spread the use of VDM++ further.

Table 11.10: COCOMO calculation drivers for options subsystem

Area	Parameter/result
DSI	60206
Mode	semidetached
Cost Drivers	
Required software reliability	normal
Database size	normal
Product complexity	normal
Execution time constraint	normal
Virtual machine volatility	normal
Computer turnaround time	low
Analyst capability	very high
Applications experience	normal
Programmer capability	high
Virtual machine experience	normal
Language experience	normal
Modern programming practices	high
Use of software tools	high
Required development schedule	very low
Result	
Effort required	147.2 person-months
Development schedule	14.3 months

Part IV

From Models to Code

12

Concurrency in VDM++

This part of the book deals with three relatively advanced topics that have a bearing on how design models are constructed and used in industry. The aim of this chapter is to demonstrate how systems with concurrency may be modelled. Chapter 13 deals with formal aspects of quality assurance in models. Chapter 14 introduces the possibility of transforming models to code, using Java as the example target language.

On completion of this chapter the reader should understand how threads can be modelled in VDM++, how multiple threads are able to communicate with each other and how it is possible to model synchronization of this communication. The chapter introduces a running example, based on the well-known POP3 protocol, which is used through the remaining chapters.

12.1 Introduction

Concurrent systems are those in which different activities may occur at the same time, whether on a single processor or on multiple processors. In either case, the use of common resources means that one activity can be affected by the actions of another, adding complexity to the task of understanding and reasoning about the system composed of these activities. Given that developing sequential systems is difficult in itself, why make the task more difficult by allowing concurrency? There are a number of answers to this question.

First, some systems are inherently concurrent. For example, in a digital flight control system, the various devices controlled by the system, such as the control surfaces, navigation and fuel management, will operate concurrently. A sequential system is not an option in this instance.

In other cases, the use of concurrency provides a better overall solution for the user. For instance, in a typical window interface to an operating system, the user would not expect all windows to freeze while one particular window completes a task; on the contrary the user expects to be able to begin a task in one window and still be able to perform tasks in other windows.

Whatever the motivation for building them, concurrent systems are here to stay. However, their increased complexity only heightens the need for the use of precise models. VDM++ is ideally suited to the construction of such models, because it supports both the modelling of concurrent activities, using *threads*, and the modelling of synchronisation across these concurrent activities. As will be demonstrated in this chapter, being able to express this synchronisation is of paramount importance when modelling concurrent systems.

This chapter begins by giving an introduction to the features of VDM++ expressly intended for modelling concurrency; following this an example of a concurrent system is described, and the construction of a VDM++ model is presented. The presentation is designed both to teach the basics of modelling concurrency in VDM++ and to highlight some of the practical issues that can occur. The concurrent system in question is a server for the POP3 protocol [RFC1939]. This is a protocol supported by all major email clients to fetch email messages from the email server. The chapter concludes with a summary of the main points.

12.2 Modelling Concurrency: Threads and Permission Predicates

In VDM++, concurrent systems can be modelled using *threads*. A thread represents an independent sequence of computations. Multiple threads may exist and be executed within one model, possibly at the same time. Threads share the same name space and communicate with each other using shared objects. To maintain the integrity of shared objects, it is often necessary to ensure that concurrent access to them is synchronised. This can be modelled using *permission predicates*. In this chapter these two features of VDM++ are explored in more detail.

12.2.1 Threads

A thread is modelled using a class, which has a thread section. For instance, the following class contains a simple thread:

```
class SimpleThread

thread
  let - = new IO().echo("Hello World!")
  in
    skip

end SimpleThread
```

This thread outputs the string "Hello World!" on standard output using VDMTools I/O standard library and then terminates. To execute the thread, a start statement should be used with an instance of SimpleThread, for example:

```
class ThreadUser

operations

UseThread: () ==> ()
UseThread() ==
  start(new SimpleThread())

end ThreadUser
```

Further examples of threads are given in Section 12.3.

12.2.2 Permission Predicates

Threads that operate in isolation are not very useful; they become much more valuable when they are able to communicate with each other. The price for this utility is the need to ensure that communication is correctly synchronised. To illustrate this, consider the following example of a system involving a single Producer thread, and a single Consumer thread:

```
class Producer

instance variables

b : Buffer

operations

Produce: () ==> seq of char
Produce() == ...

thread
  while true do
    b.Put(Produce())
end Producer
```

```
class Consumer

instance variables

b : Buffer

operations

Consume: seq of char ==> ()
Consume(d) == ...

thread
  while true do
    Consume(b.Get())
end Consumer
```

Here the `Producer` thread and the `Consumer` thread communicate with each other through one common shared instance of a one-place `Buffer` object. Figure 12.1 shows this graphically. Suppose this class is modelled as follows:

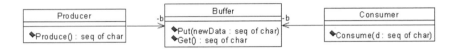

Fig. 12.1: Class diagram showing the `Producer-Consumer` system

```
class Buffer

instance variables

data : [seq of char] := nil

operations

public Put: seq of char ==> ()
Put(newData) ==
  data := newData;

public Get: () ==> seq of char
Get() ==
  let oldData = data
  in
  ( data := nil;
    return oldData
  )

end Buffer
```

What happens in this system, if the Producer thread produces data faster than the Consumer can consume it? Following the model, the Consumer will only acquire the last item that the Producer placed in the buffer; any items placed on the buffer between the Consumer's last call to Get and the Producer's last call to Put will be lost. In some systems this loss may be acceptable (e.g., for a streamed media client); however, often such a loss of data is unacceptable and should be prohibited. Note that using a sequence of data to represent a number of calls to Put instead of just one data item does not solve the problem but merely postpones it.

To model such a system, where loss of data is unacceptable, VDM++ provides *permission predicates*. These are predicates that describe when a call to an operation may be carried out. Permission predicates are described in the sync section of a class. The syntax for a permission predicate is:

```
sync

per operation name => predicate;
    . . .
```

The predicate may refer to the class's instance variables; it may also refer to a number of special variables exclusive to permission predicates, known as history counters, which provide information about the number of times each operation in the current instance of the class has been requested, activated and completed. These variables are shown in Table 12.1.

Table 12.1: History counters

Counter	Description
`#req op`	The number of times that *op* has been requested
`#act op`	The number of times that *op* has been activated
`#fin op`	The number of times that *op* has been completed
`#active op`	The number of invocations of *op* that are currently active (i.e., have been activated but not yet completed)

In the case of the `Producer-Consumer` system, to specify no loss of data it is necessary to prevent calls to `Put` being activated when the `Consumer` has not yet fetched the last data item. A simple way to express this would be the following permission predicate:

```
per Put => data = nil;
```

To understand this permission predicate, consider a simple sequence of calls. Suppose that `Put` is called when `data` is `nil`. The permission predicate is true so the call to `Put` is activated and completed. Suppose now that another call to `Put` is made. The previous call to `Put` ensured that `data <> nil`. Hence now the permission predicate is false. Therefore the thread in which this call was made (the `Producer` thread) is *blocked*; another thread will continue execution – in this case the `Consumer` thread.

Consider the opposite situation, in which the `Consumer` thread consumes data faster than the `Producer` can produce it. In this case, it is probably neither meaningful nor useful for the `Consumer` to fetch nonexistent data. Thus a permission predicate could also be used to restrict `Get`:

```
per Get => data <> nil;
```

This is not the only possible formulation of these permission predicates; for instance, the preceding permission predicate could also have been formulated as:

```
per Get => #fin(Put) - #fin(Get) = 1;
```

This states that `Put` has finished one more time than `Get`; from the permission predicate for `Put` it is known that two successive calls to `Put` are not possible, and therefore this condition implies that the buffer must contain a data item.

Exercise 12.1 Reformulate the permission predicate for `Put` using history counters. □

Thus far the problems relating to different threads operating at different rates have been resolved. However, what happens if the run-time scheduler switches out one thread in the middle of an operation and switches another thread in? If the two operations operate on shared data (as is the case with `Put` and `Get`), the result could be nondeterministic, depending on the precise run-time scheduling behaviour. In many cases this nondeterminism is undesirable. VDM++ provides a way to prevent this, using the keyword `mutex` (for mutual exclusion). Adding the line

```
mutex(Put, Get);
```

to the `sync` section of the class ensures that during execution of a call to `Put`, calls to `Get` are blocked, and vice versa.

Exercise 12.2 It is possible to reformulate any `mutex` condition as permission predicates. Reformulate `mutex(Put, Get)` as permission predicates on `Put` and `Get`, respectively. *Hint:* Use history counters. □

It is worth emphasising the difference between permission predicates and operation preconditions. As described earlier, a permission predicate is used to decide whether an operation request is activated or blocked; a precondition is used to determine whether a call to the operation is well-formed. A permission predicate evaluating false means that another thread will be executed; a false precondition can be roughly equated to a run-time failure.

Putting these conditions together, the buffer class finally takes the following form:

```
class Buffer

instance variables
  data : [seq of char] := nil

operations

public Put: seq of char ==> ()
Put(newData) ==
  data := newData;

public Get: () ==> seq of char
Get() ==
  let oldData = data
  in
  ( data := nil;
```

```
      return oldData
   )

sync
   per Put => data = nil;
   per Get => data <> nil;
   mutex (Put, Get);

end Buffer
```

This buffer works as expected if there is just one producer. However, if there are multiple producers, it is possible in principle for several producers to be activated at the same time. For instance, if one producer is activated, then that thread is switched out before the data instance variable is updated, and another producer could be activated. This is in fact a general problem that arises when an operation's permission predicate uses the object's state directly and multiple threads are competing to activate the operation. In such cases it is therefore preferable to use history counters.

In general, if all threads are blocked, i.e., all threads have reached permission predicates that are false, the system has *deadlocked*. This is a design fault and yields a run-time error. A large amount of the effort involved in designing concurrent systems is spent on avoiding deadlocks, because diagnosing the cause of a deadlock can be exceedingly difficult. Thus the effort involved in correctly modelling permission predicates should be neither underemphasised nor underestimated.

12.3 A POP3 Server

The concepts of concurrency modelling will be illustrated in VDM++ using the example of a POP3 server. POP3 is a protocol that allows email clients to fetch mail from a server. A server contains a collection of email messages for a number of users and is capable of communicating with multiple email clients simultaneously. The protocol is specified in natural language [RFC1939] and is supported by all major email clients. The purpose of the model developed here is to illustrate concurrency concepts and, as usual, it is this purpose that determines the abstraction decisions made.

12.3.1 Overview

The main class in the model is POP3Server. This communicates with the outside world using an instance of the class MessageChannelBuffer. When a client initiates communication with the server via this object, the server spawns a thread, described in the class POP3ClientHandler. This thread communicates with this client using an instance of the MessageChannel class provided by the client, until the client ends the session, at which point the thread dies. An overview of the main classes is shown in Figure 12.2, in which the principal public operations are listed.

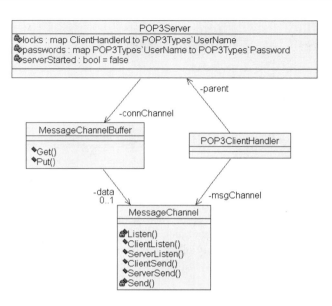

Fig. 12.2: Overview of POP3Server

12.3.2 The POP3Server Class

The class `POP3Server` uses instance variables to store three kinds of information:

`maildrop` The actual mail messages for all of the different users supported by the server.

`passwords` A password mapping, recording the password for each user name. In accordance with the model's purpose, the details of password encryption are not included.

`locks` A mapping relating the different client handlers to the user names provided by each client handler's client. Locks are described later.

```
class POP3Server
...
instance variables
  maildrop      : MailDrop;
  passwords     : map POP3Types'UserName to
                  POP3Types'Password;
  locks         : map ClientHandlerId to
                  POP3Types'UserName;
  serverStarted : bool := false;
```

```
types

public MailDrop = map POP3Types'UserName to MailBox;
public MailBox ::
  msgs    : seq of POP3Message
  locked : bool;
public ClientHandlerId = nat;
end POP3Server
```

The class `POP3Message` is an abstract representation of an email message, details of which can be found on the book's Web pages.

The overall system is inherently concurrent: At any given time there could exist many `POP3ClientHandler` threads, each with its own client, and each having access to the shared `POP3Server` object. Thus, as observed in Section 12.2.2, it is important to ensure that the integrity of the data is secure in the presence of multiple clients. This is achieved, partly due to the POP3 protocol itself and partly by the use of permission predicates.

The POP3 protocol requires that servers provide a *locking mechanism*. A client handler acquires a lock if the user name it has received from the client is not already locked. When the client handler terminates, the lock is released. This prevents the same user name being used by multiple client handlers simultaneously.

The locking mechanism is not sufficient in itself to ensure the integrity of the server's data. For instance, if two different client handlers attempted to update the instance variable `maildrop` concurrently, data could be lost or the maildrop's contents could be corrupted. On the other hand, it should not be a problem if two different client handles attempt to read `maildrop` at the same time. It is therefore necessary to use `mutex` conditions to control access to `maildrop`. To make this easier, all access to `maildrop` is performed via two private operations:

```
class POP3Server
...
operations

SetUserMessages: POP3Types'UserName * seq of POP3Message
                 ==> ()
SetUserMessages(user, newMsgs) ==
  maildrop(user) := mu(maildrop(user), msgs |-> newMsgs);
end POP3Server
```

```
class POP3Server
...
GetUserMail: POP3Types'UserName ==> MailBox
```

```
GetUserMail(user) ==
  return maildrop(user);
end POP3Server
```

Access to `maildrop` is then controlled in the following manner:

1. At any time there may only be one active invocation of `SetUserMessages`.
2. Invocations of `SetUserMessages` may not be activated if there are any active invocations of `GetUserMail`, and vice versa.
3. Any number of invocations of `GetUserMail` may be active at the same time.

Translating these requirements into VDM++ the following `sync` conditions are obtained:

```
class POP3Server
...
sync
  mutex(SetUserMessages);
  mutex(SetUserMessages, GetUserMail)
end POP3Server
```

Note that a `mutex` with just one parameter indicates that only one invocation of the operation may be active at any time.

Exercise 12.3 Access to the instance variable `locks` is achieved via the operations `AcquireLock`, `ReleaseLock` and `IsLocked`. The former two modify `locks`, while the latter just reads `locks`. Give `mutex` conditions that ensure that the integrity of `locks` is preserved. □

12.3.3 Client Handler States

A `POP3ClientHandler` accepts commands from its client and sends responses back. It effectively acts as a state machine, because certain commands are only meaningful in particular states. There are three states:

authorisation The client has not yet been authorised by the client handler. A user name/password pair has not yet been successfully submitted.
transaction The client is able to interrogate the client handler about the contents of the mail box for the client's user. The client is also able to mark messages for deletion, but messages are not actually deleted in this state.
update The client has ended the session; messages marked for deletion are actually deleted from the server, then the thread dies.

The primary functionality in the client handler occurs during the transaction state. It is here that messages may be fetched or marked for deletion.

The different states are modelled in `POP3ClientHandler` using a union of quote types:

```
class POP3ClientHandler
...
types

ServerState = <Authorization> | <Transaction> | <Update>;

end POP3ClientHandler
```

12.3.4 Starting a Client Session

The server begins a session with a client when a client initiates communication with the server using the server's connection channel. The client passes an instance of a `MessageChannel` object, which is an abstract communications channel (in practice this might be an IP address). The details of `MessageChannel` are discussed in Section 12.4.

To be available, the server listens continually for clients:

```
class POP3Server
...
thread

while true do
( let msgChannel = connChannel.Get()
  in
    CreateClientHandler(msgChannel);
  serverStarted := true;
)
end POP3Server
```

Here, `connChannel` is an instance variable:

```
class POP3Server
...
instance variables
  connChannel: MessageChannelBuffer;
end POP3Server
```

The class `MessageChannelBuffer` is just a renaming of the `Buffer` class described in Section 12.2.2, modified to allow references to `MessageChannel` to be passed. `CreateClientHandler` is an operation that spawns a new instance of `POP3ClientHandler` for this client and starts the handler's thread:

```
class POP3Server
...
CreateClientHandler: MessageChannel ==> ()
CreateClientHandler(mc) ==
  start(new POP3ClientHandler(self, mc));

end POP3Server
```

12.3.5 Interacting with the Client

Interaction between the client handler and the client is based on a dialogue consisting of `ClientCommands` and `ServerResponses`. The types `ClientCommand` and `ServerResponse` are defined in the class `POP3Types`. A `ClientCommand` may be either a standard command or an optional command (as defined in [RFC1939]):

```
class POP3Types
...
types

public ClientCommand = StandardClientCommand |
                       OptionalClientCommand;
public StandardClientCommand = QUIT | STAT | LIST | RETR |
                               DELE | NOOP | RSET;
public OptionalClientCommand = TOP | UIDL | USER | PASS |
                               APOP;

end POP3Types
```

For brevity, only the type `RETR` will be described here; the other types in these unions have very similar definitions and may be found on the book's Web pages. Note that the names of these types are presented in uppercase to be consistent with the definition of the corresponding commands in the POP3 standard.

```
class POP3Types
...
public RETR :: messageNumber : nat;
end POP3Types
```

The RETR command takes one argument: the number of the message to be retrieved.
Therefore the RETR record type has one field, corresponding to this argument. Note
that some of the POP3 commands take no arguments; these are modelled using record
types with no fields. An alternative would have been to use a quote type, but in this
case using a record type allows uniformity between the various POP3 commands.

The client handler replies with a ServerResponse value:

```
class POP3Types
...
public ServerResponse = OkResponse | ErrResponse;
public OkResponse ::  data : seq of char;
public ErrResponse :: data : seq of char;
end POP3Types
```

The ServerResponse is either positive, and then it includes any information that
the client requested, or negative, and then it includes an error message.

Top-level interaction between the client handler and the client takes place in the
client handler's thread. The client handler listens on the MessageChannel for
ClientCommands and for each such command issues a ServerResponse. This
is repeated until the client sends a QUIT command:

```
class POP3ClientHandler
...
thread
( dcl cmd: POP3Types`ClientCommand;
  id := threadid;
  cmd := msgChannel.ServerListen();
  while (cmd <> mk_POP3Types`QUIT()) do
  ( msgChannel.ServerSend(ReceiveCommand(cmd));
    cmd := msgChannel.ServerListen()
  );
  msgChannel.ServerSend(ReceiveCommand(cmd));
)
end POP3ClientHandler
```

Note here the use of the `threadid` keyword. In VDM++ `threadid` yields the unique identifier for the currently executing thread. Under VDMTools `threadid` is a dynamically generated value, unique across the execution session. It is used here to obtain a unique identifier for this session, as required by [RFC1939].

The core of the client handler is the operation `ReceiveCommand`. This is the top-level operation for handling commands received from the client:

```
class POP3ClientHandler
...
ReceiveCommand: POP3Types'ClientCommand
                ==> POP3Types'ServerResponse
ReceiveCommand(c) ==
    cases c:
        mk_POP3Types'QUIT()      -> ReceiveQUIT(c),
        mk_POP3Types'STAT()      -> ReceiveSTAT(c),
        mk_POP3Types'LIST(-)     -> ReceiveLIST(c),
        mk_POP3Types'RETR(-)     -> ReceiveRETR(c),
        mk_POP3Types'DELE(-)     -> ReceiveDELE(c),
        mk_POP3Types'NOOP()      -> ReceiveNOOP(c),
        mk_POP3Types'RSET()      -> ReceiveRSET(c),
        mk_POP3Types'TOP(-,-)    -> ReceiveTOP(c),
        mk_POP3Types'UIDL(-)     -> ReceiveUIDL(c),
        mk_POP3Types'USER(-)     -> ReceiveUSER(c),
        mk_POP3Types'PASS(-)     -> ReceivePASS(c)
    end

end POP3ClientHandler
```

Here, the type of the command received is examined and then control is delegated to an operation dedicated to that command. To be able to compute a response, the following instance variables are used: `ss`, the current server state; `parent`, a reference to its `POP3Server` parent; and `user`, the user name provided by the current client (if any user name has been provided):

```
class POP3ClientHandler
...
instance variables
  ss     : ServerState;
  parent: POP3Server;
  user   : POP3Types'UserName;
end POP3ClientHandler
```

As an example of how a command is dealt with, consider the operation, which receives a RETR command:

```
class POP3ClientHandler
...
operations

ReceiveRETR: POP3Types'RETR ==> POP3Types'ServerResponse
ReceiveRETR(retr) ==
  if ss = <Transaction>
  then
    if parent.IsValidMessageNumber(user,
                                   retr.messageNumber)
    then let msgText = parent.GetMessageText(
                            user,
                            retr.messageNumber),
             sizeText =
               int2string(parent.GetMessageSize(
                            user,
                            retr.messageNumber))
         in
             return mk_POP3Types'OkResponse(sizeText ^
                                            "\n" ^
                                            msgText)
    else return
           mk_POP3Types'ErrResponse(unknownMessageMsg)
  else return mk_POP3Types'ErrResponse(negativeStatusMsg)
end POP3ClientHandler
```

This operation first checks that the client handler is in the appropriate state for this command; if it is, it uses the instance variable parent (a reference to its POP3Server parent) to check that the message number included in the command refers to a non-deleted message for the user whose name is given by the instance variable user; this is known as a valid message.

If the message number is valid, then the client handler responds with the size of the message followed by the message text. Here int2string is an auxiliary function used to convert an integer to its string representation. Its definition is straightforward and therefore omitted. The different message strings used in the responses are defined as constant values in the class:

```
class POP3ClientHandler
...
values

unknownMessageMsg: seq of char = "No such message";
negativeStatusMsg: seq of char =
                   "Wrong state for this command";
end POP3ClientHandler
```

The remaining operations for receiving commands from the client are defined in a similar manner to `ReceiveRETR` and are therefore not described further.

12.4 Synchronising Communication in the POP3 Server

The client handler and client communicate using a shared `MessageChannel` object. Therefore it is important that they synchronise correctly during communication; otherwise messages could be lost. As the client handler reacts to commands from the client, the sequence of activations should follow that indicated in Figure 12.3.

The basic behaviour of the `MessageChannel` is therefore straightforward. First, it will have an instance variable representing the command being sent or the response received:

```
class MessageChannel
...
instance variables
  data : [POP3Types`ClientCommand |
          POP3Types`ServerResponse]   := nil;
end MessageChannel
```

There are then four public operations, `ClientSend`, `ServerListen`, `ServerSend` and `ClientListen`, corresponding to the four different activations shown in Figure 12.3. These are defined in terms of the private operations `Send` and `Listen`:

```
class MessageChannel
...
operations

Send: POP3Types`ClientCommand |
      POP3Types`ServerResponse ==> ()
Send(msg) ==
  data := msg;
```

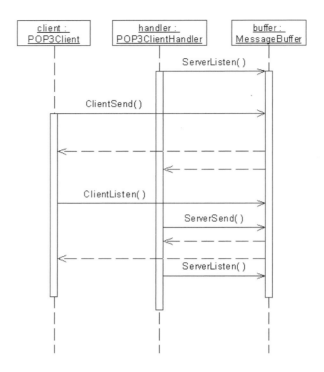

Fig. 12.3: Sequence of activations in `MessageChannel`

```
Listen: () ==> POP3Types'ClientCommand |
               POP3Types'ServerResponse
Listen() ==
let d = data in
  ( data := nil; return d
  );
end MessageChannel
```

For example, `ServerSend` uses `Send` to put data on the channel:

```
class MessageChannel
...
operations

public ServerSend: POP3Types'ServerResponse ==> ()
```

```
ServerSend(p)  ==
  Send(p);
end MessageChannel
```

The other three operations are defined similarly.

The challenge here is to model synchronization in such a way that the sequencing described in Figure 12.3 is observed. Each of these four operations will be given a permission predicate; here two of them are considered.

The permission predicate for ClientSend is the easiest to understand. The operation ClientSend initiates a communications round, so it may only be activated if the number of completions of all four operations is identical. This gives the following permission predicate:

```
class MessageChannel
. . .
  per ClientSend => #fin(ServerSend)  = #fin(ClientListen)
                    and
                    #fin(ClientSend)  = #fin(ServerListen)
                    and
                    #fin(ServerSend)  = #fin(ClientSend)  ;
end MessageChannel
```

In the case of ServerListen, after completion of ClientSend, it must be the case that #fin(ClientSend) = #fin(ServerListen) + 1 (follows from the permission predicate for ClientSend). Further inspection reveals that this is the only time in the round that this is the case. This yields the following permission predicate:

```
class MessageChannel
. . .
  per ServerListen => #fin(ClientSend) - 1 =
                      #fin(ServerListen);
end MessageChannel
```

Exercise 12.4 Formulate a permission predicate for ServerSend by considering the number of completions of the operations ClientSend, ServerListen and ServerSend. □

Exercise 12.5 Formulate a permission predicate for ClientListen. □

One point here is worthy of note: The operations ClientSend and ServerSend, and ClientListen and ServerListen are functionally equivalent to, respectively, Send and Listen. Why, therefore, have this extra layer of operations? The answer is that it would be less easy to distinguish calls from the client and the client handler without this extra layer. This would lead to much more complicated permission predicates, which in turn would increase the risk of the model having a run-time deadlock. In general, in the development of concurrent models it is not unusual to introduce an extra layer of operations in this manner, whose main purpose is to allow the straightforward modelling of synchronisation.

12.5 Summary

This chapter has concentrated on how to model concurrent systems using VDM++. The use of concurrency allows models to be developed in which new paradigms can be used, for example, event-driven systems and client-server systems. However, as has been illustrated by the examples given, concurrent systems are highly vulnerable to subtle design errors.

VDM++ is helpful for modelling concurrent systems for two reasons:

1. The different sequential activities in the system that combine to give a concurrent system can be modelled using *threads*. All of the powerful abstraction features described in the earlier chapters are available when using threads, so the individual sequential behaviours can be precisely captured and recorded. This was demonstrated by the class POP3ClientHandler.

2. Synchronisation between different threads is possible using *permission predicates*. As demonstrated using both the Producer-Consumer example and the class MessageChannel, giving permission predicates is a necessity if communication between threads is to follow any rules. In the absence of permission predicates such communication would be chaotic. Permission predicates are typically expressed using mutex or in terms of history counters.

Exercise 12.6 Suppose in the Producer-Consumer problem that two Producers and Consumer all shared the same instance of Buffer. Redesign Buffer to cope with this situation. □

13

Model Quality

This chapter demonstrates specific techniques that can be used for assessing and improving model quality. The POP3 example introduced in Chapter 12 forms the basis of the main example. Some knowledge of CORBA and associated technologies, as well as the Java [Gosling&00] programming language is assumed, but it is not essential to grasp the principles covered.

13.1 Introduction

How good is a model? The concept of quality is often treated vaguely but the ISO8402 standard [ISO8402] provides a working definition:

> the totality of features and characteristics of a product, process or service that bear on its ability to satisfy stated or implied needs.

A model's quality can therefore be seen as its "fitness for purpose." Recall from Chapter 2 that it is vital to keep the model's purpose foremost when making the abstraction decisions. The same is true when assessing its quality. Models are developed for a wide range of purposes: VDM and VDM++ have been applied to tasks ranging from the expression of fundamental ideas in computing science through to detailed hardware specification. All these applications have their own quality criteria. A model developed for theoretical analysis needs to be clear, concise and expressed using a notation that encourages insight. A model of a detailed software design needs to be susceptible to machine-assisted analysis to ensure that it expresses the properties of the system under construction, is capable of being tested, and has a good clear basis for deriving code. This chapter is about the techniques available for assessing whether a model has the qualities required of it, the costs and benefits of the techniques, and the tool support available for them.

Using a formal notation such as VDM++ to enhance an object-oriented model widens the range of qualities that are capable of being tested rather than merely argued over. For example, instead of arguing about internal logical consistency it becomes possible to verify some aspects of consistency, such as syntax and type correctness, to a high level of confidence and largely automatically. Some properties, for example, ensuring that operations respect invariant properties, can be partially, if not completely, automated, and still be reasoned about rigorously. Of course, some qualities, such as whether a model is comprehensible, remain beyond the scope of formality. This chapter concentrates on the qualities that can be better assessed using the enhanced rigour of an object-oriented and VDM++ model.

Model quality can be considered in two different but complementary ways: the internal consistency requirements. Internal consistency covers those qualities internal to the model that ensure that the model does not contradict itself – "does the model describe something?" External consistency covers the model's correctness or faithfulness to the requirements – "does the model describe the right thing?" One of the advantages of using VDM++ is that the model is described abstractly yet precisely. However, there are no guarantees that the model actually describes the desired system. This issue is not a new one: Ascertaining the correctness and completeness of user requirements has been a problem occupying software engineers for many years.

There are several different approaches to assessing internal consistency; syntax and type checking (Sections 3.4.2 and 3.4.4) can be considered simple forms of internal consistency checking. A more advanced form is integrity checking, an analysis technique that identifies potential sources of run-time errors in models, described in Section 13.2.1.

Assessing external consistency is not an exact science because requirements are typically formulated using ambiguous natural language. Because it can never be said with complete certainty that a model correctly captures the users' requirements, the issue is therefore one of increasing confidence. Recall that this is the task of *validation*, introduced in Chapter 1 as "the process of examining a product to determine conformity with user needs." Validation is crucially important because it is not usually the users themselves who create the model. Instead the domain experts, rarely themselves expert in modelling, communicate their requirements to the modellers. This gap is a rich source of potential error, misunderstanding and misinterpretation. Validation assesses the degree to which the modeller has been able to capture the needs of the domain experts. Failure to catch errors at the design stage can lead to expensive error correction much later. An ideal approach to validation would entail domain experts inspecting the model themselves. Unfortunately, domain experts are rarely fluent in the modelling languages used by computer scientists. However, where the model is executable, it is possible for them to assess the model by executing it and analysing its behaviour, then communicating potential improvements back to the modeller. This approach to animating models for validation is made possible using the application programmer's interface (API) available in VDMTools, described in Section 13.3.

13.2 Assessing Internal Consistency: Integrity Checking

Internal consistency checking is primarily about identifying potential errors in a model by pointing out areas in which there may be some anomaly: an inconsistency or potential undefinedness, for example. A large catalogue of integrity checks exists, each aiming to identify a particular potential problem.

The very simplest integrity checks – syntax conformance and simple type checks – are completely automated, as they are with any high-level programming language compiler. However, as any programmer knows, these checks offer only a very limited increase in confidence that the model meets its requirements!

As an example, consider a very simple model handling the booking of long-distance trips. A trip is modelled as a sequence of flights, each with a departure airport and a destination airport:

```
class Trip

types
Flight :: departure   : seq of char
          destination : seq of char

instance variables
  journey: seq of Flight;

inv forall i in set {1,...,len journey -1} &
      journey(i).destination = journey(i+1).departure

operations

AddFlight: Flight ==> ()
AddFlight(f) ==
  journey := journey ^ [f]

end Trip
```

The operation `AddFlight` updates the `journey` by appending the new flight on to the tail end using a perfectly valid sequence operator. The model would pass a static type check. However, the modeller has failed to take account of the invariant in that the operation could add a new flight that breaks the invariant; the journey would cease to be valid. A simple static type check fails to find this potential difficulty.

In general, static type checking is simply structural. It involves checking that the types of values generated or modified by operations have the correct "shape," i.e., are numbers, sequences, records with the right fields, etc. Where invariants are present, the check is more subtle. In the flight booking example, it would be necessary to verify that the `AddFlight` operation could never be called in a way that would lead to violation of the invariant. This example has a fairly simple invariant: If very expressive invariants are permitted, it is not possible to build a tool that can do the check entirely automatically in all circumstances. However, invariant preservation is one example of the checks that can be performed with machine assistance because a formal model, rather than an informal one, is available. Section 13.2.1 presents a range of example integrity properties at increasing levels of sophistication. The examples are drawn from the models developed so far in the book.

13.2.1 Machine-Assisted Checking: Formal Integrity Properties

Protection of Partial Operators

Modelling languages, like many programming languages, can include partial operators, which may not deliver defined results for all possible arguments. Partial operators in VDM++ were introduced in Section 4.4.1 and subsequently. Examples include numerical division, where division by zero is undefined, and the hd operator, undefined if applied to the empty sequence. The application of a partial operator is therefore to be treated with care: points at which such operators are used should be highlighted during internal consistency checking.

An interesting example occurs in the POP3 server model developed in Chapter 12, specifically in the class POP3Server.

One of the instance variables, MailDrop, is a mapping from mail users to their mailboxes:

```
class POP3Server
...
instance variables
  maildrop       : MailDrop;
  ...

types

public MailDrop = map POP3Types'UserName to MailBox;
public MailBox ::
  msgs    : seq of POP3Message
  locked : bool;
...
end POP3Server
```

The operation GetUserMail simply returns the mailbox for a given user:

```
class POP3Server
...
GetUserMail: POP3Types'UserName ==> MailBox
GetUserMail(user) ==
  return maildrop(user);
```

Notice that the definition of GetUserMail uses a partial operator, namely a mapping application where the user is looked up in the mailbox mapping. The relevant integrity check here entails showing that, in this context, the operation is defined. A validated version of the POP3Server class includes a precondition to ensure that the map application is sound:

```
class POP3Server
...
GetUserMail: POP3Types'UserName ==> MailBox
GetUserMail(user) ==
   return maildrop(user)
pre UserKnown(user);
end POP3Server
```

The precondition uses an operation that checks that the user is known in the maildrop. It records a restriction on the domain of GetUserMail, so this operation must be regarded as partial, and this character has to be taken into account wherever the operation is used. For example, the GetUserMessages operation returns the sequence of messages in the msgs component of the mailbox:

```
class POP3Server
...
GetUserMessages: POP3Types'UserName ==>
                  seq of POP3Message
GetUserMessages(user) ==
   return GetUserMail(user).msgs;
```

This too needs to deal with the partial character of GetUserMail and, again, a precondition can be used:

```
GetUserMessages: POP3Types'UserName ==>
                  seq of POP3Message
GetUserMessages(user) ==
   return GetUserMail(user).msgs
pre UserKnown(user);
```

In turn, the GetUserMessages operation is used in still other operations, for example, in RemoveDeletedMessages, and once again the precondition could be propagated outward:

```
class POP3Server
...
public RemoveDeletedMessages: POP3Types'UserName ==>
                                        bool
RemoveDeletedMessages(user) ==
let oldMsgs = GetUserMessages(user),
    newMsgs = [oldMsgs(i) | i in set inds oldMsgs
                            & not oldMsgs(i).IsDeleted()]
in
  (SetUserMessages(user, newMsgs);
   return true
  )
pre UserKnown(user);
```

If the precondition is omitted, the reader has to trace back through several levels and know about the protocol to gain confidence that `GetUserMail` is called safely. Although the calls may be sound within the model, how confident can one be that future modifications to the model will respect the partial operator? In general, it is good to record the conditions under which a function is defined by using a precondition.

An alternative to using preconditions is to build some explicit error handling into the function or operation, rendering the operation itself total. This is often done using an optional return type, so that `nil` can denote the error value. Such an approach to the `GetUserMail` operation would result in the following definition:

```
GetUserMail: POP3Types'UserName ==> [MailBox]
GetUserMail(user) ==
  return if user in set dom maildrop
         then maildrop(user)
         else nil;
```

Using this definition, `GetUserMail` is total, but every invocation of the operation must still take account of the possibility of `nil` being returned. A further alternative is to use any built-in exception handling features of the modelling language, such as the VDM++ exit statement. Both of these mechanisms allow operations to be executed without separate precondition checking, but this comes at a price. Preconditions serve a very valuable purpose by making explicit in the model the conditions under which functions and operations should be applied. Trying to work this information out from a model that uses error values or exits is much more difficult.

Invariant Preservation

One of the most significant integrity checks is that of invariant preservation. Any function must ensure that the result is not just structurally type-correct, but that it also respects the invariant on the result type. Similarly operations must ensure that invariants on instance variables and any result types are respected. Formally, this invariant preservation is required to hold for any inputs that respect the function or operation's precondition.

The small flight booking system example at the start of Section 13.2 illustrates this. The `AddFlight` operation has the potential to change the `journey` instance variable in such a way as to break the invariant condition that the flights are linked. The problem could, once again, be resolved using a precondition:

```
class Trip
...

AddFlight: Flight ==> ()
AddFlight(f) ==
   journey := journey ^ [f]
pre journey(len journey).destination = f.departure
```

Look carefully at the precondition, remembering to consider any possible applications of partial operators. The sequence lookup `journey(len journey)` could be undefined if `journey` is empty and the length is 0 (VDM++ sequence indexes start at 1). The precondition should take account of this:

```
class Trip
...

AddFlight: Flight ==> ()
AddFlight(f) ==
   journey := journey ^ [f]
pre journey <> [] =>
      journey(len journey).destination = f.departure
```

Satisfiability

Much of the discussion of internal consistency checking so far has concentrated on explicit function and operation definitions. Implicit definitions, though comparatively rare in executable models, have a consistency check, too, in the form of the *satisfiability* check. A pre/post specification is satisfiable, if, for all states and inputs satisfying the precondition, there exists some output and "after" state satisfying the postcondition and any relevant invariants.

For an example, recall the robot controller from Chapter 6. A route for the robot is modelled as a set of points. An invariant records some consistency constraints on routes:

```
class Route
...
instance variables

points: set of Point;
end Route
```

```
class Route
...

inv forall p1, p2 in set points &
       p1.GetCoord() = p2.GetCoord() => p1 = p2 and
    forall p in set points &
      p.GetIndex() <> card points
      => GetNext(p).GetCoord() in set
          {n.GetCoord() | n in set p.Neighbour()}
end Route
```

The function AvoidanceRoutes takes a set of obstacles to be avoided, the currentPosition of the robot and the nextWaypoint to be visited. It returns the set of obstacle-free routes to the nextWaypoint. The postcondition describes the set to be returned:

```
class Route
...
functions

static
public AvoidanceRoutes(
                obstacles: set of (nat * nat),
                currentPosition: Point,
                nextWaypoint: Point) routes: set of Route
post forall r in set routes &
       r.GetFirst().GetCoord() =
       currentPosition.GetCoord() and
       r.GetLast().GetCoord() =
       nextWaypoint.GetCoord() and
       r.GetCoords() inter obstacles = {};
end Route
```

The integrity property for this function can be stated formally as follows:

```
forall obstacles: set of (nat * nat),
        currentPosition: Point,
        nextWaypoint: Point &
  exists routes: set of Route &
     post-AvoidanceRoutes(obstances, currentPosition,
                                 nextWaypoint, routes)
```

This formal statement asserts that for all inputs there exists a result of the correct type. The "exists routes: set of Route ..." part is, course, very hard to verify automatically. In fact, a mathematical argument or proof is required in this case, and that is beyond the scope of this book.

13.3 Assessing External Consistency Using an API

An application programmer interface for a modelling support tool, such as the VDM-Tools API, provides a mechanism whereby external applications can access the functionality provided by the tools. In the context of external consistency checking, this opens up the possibility of building a simple, perhaps graphical, interface, allowing domain experts to interact with the model in a way they understand and can relate to, without having detailed knowledge of the model itself.

The VDMTools API uses CORBA to expose the tools' functionality to external applications. CORBA'S interface description language (IDL) is used to give a description of each component. In addition, an IDL description is provided for the values such as sets and sequences used within the model.

The rest of this chapter assumes the use of Java; the object request broker (ORB) used is that distributed by Sun Microsystems with the Java 2 platform. A jar library containing a Java implementation of the IDL description of the VDMTools API is distributed with VDMTools. The API classes referred to are included in this jar library. A reference guide to the use of the API is available in [APIMan]. In this section the POP3 example from Chapter 12 is used to demonstrate the details of how to use the API to assess external consistency. Before that, a brief introduction to how to use the API is given. The full set of files can be found on the book's supporting Web site.

13.3.1 Introduction to the API

Suppose the API is to be used to syntax, type check and execute the model shown in Figure 13.1.

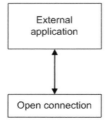

Fig. 13.1: Overview of API

```
class A
...
public op: int ==> int
op(n) ==
  return n + 1
pre n > 0
end A
```

The life cycle of an application that interacts with the API is shown in Figure 13.2.

Fig. 13.2: Life cycle of interaction with API

13.3.2 Connecting to the API

To connect to the API, the class `ToolboxClient` should be used. It provides the method `getVDMApplication`, which takes two parameters: an array of strings to pass as arguments to the tools and a flag indicating whether a connection should be made to a VDM-SL or VDM++ Toolbox. (The VDM-SL Toolbox is the sister product to VDMTools, supporting the notation of VDM-SL described in [Fitzgerald&98].) The method returns a CORBA object containing a reference to an executing tool set; if no tool set is running, an exception is thrown. This reference then needs to be *narrowed* (cast) to an instance of `VDMApplication`. In Java, if app is a field of type `VDMApplication` this might be written as

```
ToolboxClient toolboxClient = new ToolboxClient();
org.omg.CORBA.Object obj =
   toolboxClient.getVDMApplication(new String[]{},
                                 ToolType.PP_TOOLBOX);
app = VDMApplicationHelper.narrow(obj);
```

Having acquired this reference to the tools, a couple of initialisation commands must be executed. First, this client needs to register with VDMTools, because it is possible to have multiple clients simultaneously connect to a running tool; second, this client needs to call the PushTag method. This is used by VDMTools for internal house-keeping. If client is a field of type short, this might be coded as

```
client = app.Register();
app.PushTag(client);
```

It is now possible for the client to access all of the functionality available in the tools.

13.3.3 Interacting with the API

From the VDMApplication class, it is possible to access the tool features. Typically a client would begin by loading a model. This could be done by either directly opening a project using the VDMProject class or creating a new project and then adding files to this new project. This latter approach is now demonstrated.

First, a new project is created, using VDMProject:

```
VDMProject prj = app.GetProject();
prj.New();
```

Next, the files are parsed and then added to the project:

```
String path = "/local/vdm++book/validation";
String[] modelFiles = {"A.vpp"};
VDMParser parser = app.GetParser();

for (int i = 0; i < modelFiles.length; i++)
{
    String filename = path + "/" + modelFiles[i];
    prj.AddFile(filename);
    parser.Parse(filename);
}
```

Once the model has been loaded, interaction with the tools may proceed. For instance, all of the classes in a model can be type-checked. To do this, first the list of classes in the project is obtained, and then this is given to the type checker. This is demonstrated in the method `typeCheck`:

```
private void typeCheck() throws APIError {
    ModuleListHolder moduleList = new ModuleListHolder();
    app.GetProject().GetModules(moduleList);
    app.GetTypeChecker().TypeCheckList(moduleList.value);
}
```

Notice here that *holder classes* are used to get results by reference, in this case illustrated by the use of the class `ModuleListHolder`.

To execute the model, first the interpreter needs to be initialised, and then the base objects for the execution need to be created (corresponding to the use of the `create` command in the interpreter). This might be achieved as follows:

```
VDMInterpreter interp = app.GetInterpreter();
interp.Initialize ();
interp.EvalCmd("create a := new A()");
```

Having created an object, it can be used for execution. The `VDMInterpreter` class provides a number of different ways of doing this. One way would be to use the `EvalCmd` method; another way is to use the `Apply` method in the class `VDMInterpreter`:

```
try {
    VDMFactory fact     = app.GetVDMFactory();
    VDMSequence args     = fact.MkSequence(client);
    VDMNumeric intValue = fact.MkNumeric(client, 5);
    args.ImpAppend(intValue);
    VDMGeneric result =
            interp.Apply(client, "a.op", args);
    System.out.println("Result is " + result.ToAscii());
} catch (APIError e) {
    System.err.println("Unable to validate model");
}
```

Here, `Apply` takes three parameters: the client identifier, a string representing the name of the function or operation to be executed (including the object reference) and a sequence of arguments to be given to the function or operation. This sequence of arguments is a CORBA representation of a VDM++ sequence. CORBA representations of VDM++ values can be created using the `VDMFactory` class. In the preceding example the methods `MkSequence` and `MkNumeric` have been used to create, respectively, a CORBA representation of a VDM++ sequence and a CORBA representation of a VDM++ number. The result of `Apply` is a reference to a `VDMGeneric` object. The class `VDMGeneric` is the superclass of all CORBA representations of VDM++ values. This could either be narrowed, based on the known return type of the function or operation, or (as earlier) the result can be directly converted to a string.

Note here that `Apply`, like many other methods of the API, can throw the exception `APIError`. The call to `EvalCmd` must therefore be enclosed within a try/catch block.

Exercise 13.1 The method `DynPreCheck` defined in the `VDMInterpreter` class is used to control dynamic invariant checking. It takes a boolean value used to indicate whether checking should be enabled or disabled. Define a Java method called `validateOp` which takes a parameter of type `int` and executes A`op with dynamic precondition checking enabled. The method should return the result of A`op applied to the `int` parameter. □

13.3.4 Closing the Connection

When the client has finished interacting with the tools, it is good practice to close the connection cleanly. First, for the earlier call to the `PushTag` method there is a call to `DestroyTag`; this ensures that all temporary objects created specifically for this client can now be destroyed. Second, the client "unregisters" itself, so that VDMTools knows the client's session is complete.

Exercise 13.2 Write the void Java method `shutdown`, which takes no parameters and cleanly closes the connection to VDMTools. □

13.4 Validating the POP3 Example through the API

In this section the example of the POP3 server described in Chapter 12 is revisited; the model is validated by constructing a graphical user interface, through which a user can interact with the model, without necessarily having knowledge of VDM++.

It is useful to bear in mind a couple of general design principles when performing this kind of validation. First, it is important that communication with the tools is, as far as possible, isolated to a few methods. These methods may themselves be used extensively. This approach greatly eases development, debugging and maintenance.

Second, it is desirable that the dialogue between the client and the executing tool can be viewed. This is useful for both debugging purposes and making explicit the fact that validation of a model is taking place, rather than debugging an application.

In terms of the POP3 model, because a large part of the model is concerned with concurrency, it is important that the validation allows multiple clients to interact with the same model.

Finally, as will be seen in Chapter 14, intelligent use of abstract classes can allow the reuse of the GUI with the final application.

13.4.1 Overview

The POP3 client essentially consists of two Java classes: the GUI and a class used to interface between the GUI and the model. This arrangement is shown in Figure 13.3.

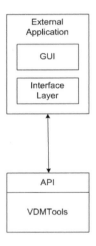

Fig. 13.3: Structure of POP3 validation application

The visual appearance of the GUI is shown in Figure 13.4, but of main interest here is the interface layer. This is explored in the next two sections, where first the issue of a single client interacting with the model is described, and after that the additional issues relating to multiple concurrent clients are examined.

13.4.2 Interacting with One Client

In the first instance it is simplest to disregard the issue of concurrency and concentrate on validation with a single client. Two classes are defined for this purpose: the GUI class `Pop3ApiGui` and the interface layer `Pop3ApiLayer`.

First, `Pop3ApiLayer` defines a number of fields:

`VDMApplication app` This is a CORBA reference to the running VDMTools instance.

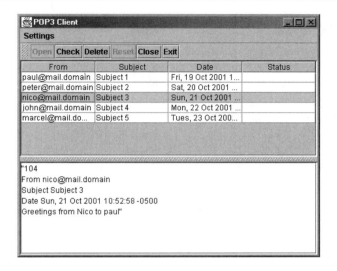

Fig. 13.4: POP3 client interface

VDMInterpreter interp This is a CORBA reference to the interpreter in the
   instance of the VDMTools to which app refers.
short client The identifier number allocated by the VDMTools to this client.
JTextArea logArea A reference to a Java swing text area object, used to show
   the dialogue between the client and VDMTools (following the design principles
   described earlier).
String channel The name of the channel that will be used within the interpreter
   for communication with the server.

Following the design principles given earlier, it is sensible to define some methods
to isolate communication between the interface layer and the API. Because a large portion of the interaction will use the method VDMInterpreter.EvalCmd, a method
is defined in Pop3ApiLayer to shadow this method:

```
private void EvalCmd(String cmd) throws APIError
{
  Log("EvalCmd(" + cmd + ")");
  interp.EvalCmd(cmd);
}
```

This method first calls the method Log to record this dialogue and then uses the class'
interp field to evaluate the command cmd in the VDMTools interpreter. The Log
method just appends the String it receives to logArea.

13.4.3 Creating VDM++ Objects

An example of the use of the `EvalCmd` method can be seen in the method called `initInterpreter`. This method initialises the interpreter, then creates the VDM++ objects needed to execute the POP3 server model:

```
private void initInterpreter() throws APIError
{
  // Ensure echoing in interpreter
  interp.Verbose(true);

  // Enable precondition checking during execution
  interp.DynPreCheck(true);
  interp.Initialize ();

  EvalCmd("create ch := new MessageChannelBuffer()");
  EvalCmd("create pt := new POP3Test()");
  EvalCmd("create server := " +
          "new POP3Server(pt.MakeMailDrop(), " +
                          "ch,pt.MakePasswordMap())");
  EvalCmd("debug pt.StartServer(server)");
}
```

Notice here that the strings given as arguments to `EvalCmd` correspond exactly to what would be typed by the user in the interpreter if direct execution of the model was taking place.

13.4.4 Interaction between Client and Server

Having created these objects within the interpreter, it is now possible for the client to begin interaction with the model. Recall from Chapter 12 that a client initiates a session with the POP3 server by creating a `MessageChannel` instance, sending this to the server, and then initiating a user/password dialogue.

The method for performing this task is called `openServerConnection`. It takes three parameters:

`String username` username to be given to the server;
`String password` password to be given to the server;
`StringBuffer response` StringBuffer in which to place any messages from the server.

The method returns true if the session has been successfully opened and false otherwise. The method is defined as follows:

```
public boolean openServerConnection(String username,
                                    String password,
                                    StringBuffer
                                       response)
{
  try {
    EvalCmd("create " + channel + " :=
            new MessageChannel()");
    EvalCmd("debug ch.Put(" + channel + ")");
    boolean status = executeCommand(
                   "USER",
                   new String[]{"\""+username+"\""},
                   response);
    if (!status)
      return false;
    status = executeCommand(
               "PASS",
               new String[]{"\""+password+"\""},
               response);
    return status;
  } catch (Exception e) {
      e.printStackTrace(System.err);
      return false;
  }
}
```

Most of the method's work is delegated to the auxiliary method `executeCommand`; this method constructs the value to be sent to the server, then decodes the response received from the server. It is instructive to consider this method in detail, as the approach to interaction described here could easily be reused in other scenarios.

This method takes three parameters:

`String title` the title of the command;
`String[] args` any arguments for the command;
`StringBuffer response` a `StringBuffer` to contain the response.

The method returns true if the response was satisfactory and false otherwise. This method itself uses three further auxiliary methods: `makeCommand`, which constructs a string corresponding to a value from the class `POP3Types`; `sendCommandResponse`, which sends a command to the POP3 server and returns the corresponding response and `checkResponse`, which checks the response received from the server:

```
private boolean executeCommand(String title,
                               String[] args,
                               StringBuffer response)
{
  String command = makeCommand(title, args);
  try {
    VDMRecord responseObj =
                sendCommandResponse(command);
    boolean status = checkResponse(responseObj);
    response.append(responseObj.GetField(1).ToAscii());
    return status;
  } catch (Exception e) {
    System.err.println("executeCommand: " +
                       e.toString());
    return false;
  }
}
```

The method makeCommand constructs a command from the record type name given by the parameter cmd and the arguments to the type constructor given by the array parameter args. The resultant string represents a value of type ClientCommand from the POP3Types class:

```
private String makeCommand(String cmd, String[] args)
{
  StringBuffer command = new StringBuffer();
  command.append("mk_POP3Types'");
  command.append(cmd);
  command.append("(");
  for (int index = 0; index < args.length; index++)
  {
    command.append(args[index]);
    if (index != args.length-1)
      command.append(",");
  }
  command.append(")");
  return command.toString();
}
```

This method explicitly creates a record value using the mk record constructor. However, no attempt is made to check that the number of arguments corresponds to the number of fields defined for this type or that the arguments are of a suitable type. This checking can be left to VDMTools.

sendCommandResponse is another auxiliary method. It sends a command to the server using the `MessageChannel`ClientSend` method and fetches the response using `MessageChannel`ClientListen`:

```
private VDMRecord sendCommandResponse(String command)
{
    VDMRecord responseRecord = null;
    try {
        EvalCmd("debug " + channel + ".ClientSend(" +
                command + ")");
        VDMFactory fact     = app.GetVDMFactory();
        VDMSequence args     = fact.MkSequence(client);
        VDMGeneric response = interp.Apply(client,channel+
                                            ".ClientListen",
                                            args);
        responseRecord = VDMRecordHelper.narrow(response);
    } catch (APIError e) {
        System.out.println("sendCommandResponse: " +
                            e.msg.toString());
    }
    return responseRecord;
}
```

Exercise 13.3 Why does `sendCommandResponse` use the `Apply` method of `VDMInterpreter` instead of the `EvalCmd` method when it fetches a response from the server? □

Note that when the response is first received from the server, it is a CORBA object of type `VDMGeneric`. However, because it is known from the model that the return type of `ClientListen` is `POP3Types`ClientCommand`, which must therefore be a record, the response may be safely narrowed to a `VDMRecord` object.

Finally, consider `checkResponse`. This auxiliary method checks that a response received from the server is a record of type `POP3Types`OkResponse`:

```
private boolean checkResponse(VDMRecord response)
{
    try {
        Log("response is " + response.ToAscii());
        return response.GetTag().equals(
                "POP3Types`OkResponse");
    } catch (APIError e) {
        System.err.println(e.toString());
        return false;
```

```
        }
    }
```

Exercise 13.4 If the return type of `MessageChannel'ClientListen` had instead been `[POP3Types'ClientCommand]` (where a nil value indicates an error of some kind), the parameter to `checkResponse` would have been a CORBA object of type `VDMGeneric`. Modify `checkResponse` to handle this situation. □

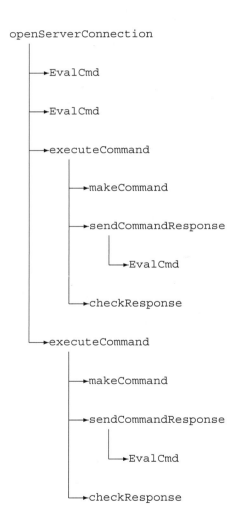

Fig. 13.5: Structure of `openServerConnection` method

13.4.5 Review

Having presented all of the details of a single interaction between the client and the server, it is sensible to review the steps taken. To this end a diagram showing the call structure of the method openServerConnection is shown in Figure 13.5. In this diagram an arrow from method a to method b indicates that method b is called from within method a.

The remaining interaction between the client and the server follows exactly the same pattern as that of openServerConnection. Therefore these are not described in more detail here.

With this client it is possible to validate the basic functionality of the POP3 server: User/password authentication, the fetching of mail, deletion of mail, resetting of message status and so on can all be validated. Moreover this validation can occur without the user *ever* having to see the details of the underlying VDM++ model. This reveals an extremely powerful technique, because it means that it is possible to communicate the model to domain experts who need not understand the details of the notation in which the model is expressed.

13.4.6 Closing the POP3 Session

When the user has finished interacting with the client and presses the "Close" button in the interface, the client should close this POP3 session. This means informing the server that the session is over and tidying up any objects created for this session. Note that this is different from the client closing the connection with the API, which occurs when the "Exit" button is clicked.

To close this POP3 session, the QUIT command should be sent to the server. Then the single object created for this session – the channel – should be destroyed in the interpreter. Destruction of this object is not strictly necessary, but it is good practice.

The method for closing the session is closeAction:

```
public void closeAction()
{
  StringBuffer response = new StringBuffer();
  executeCommand("QUIT", new String[]{}, response);
  try {
    EvalCmd("destroy " + channel);
  } catch(APIError e){
    System.out.println(e.msg);
    e.printStackTrace(System.err);
  }
}
```

Exercise 13.5 A key concern when designing an application is the modifiability of the design. In this case it would be desirable if the GUI application was resilient to changes in the underlying VDM++ model. What would be the implications for the GUI application under each of the following scenarios:

1. In the `MessageChannel` class the `ClientListen` operation is renamed to `ClientWait`.
2. Instead of being a union of record types, `ClientCommand` is respecified as an abstract base class with `QUIT`, `STAT` and other subclasses of `ClientCommand`.

□

13.4.7 Interacting with Multiple Clients

Having enabled interaction between a single client and the POP3 server, the extension to multiple clients requires resolution of two problems:

- The objects created by a client for a single POP3 session need to be unique for that client: otherwise it would be possible for one client to interfere with another client's session.
- Only the very first client that registers with the tool set should load the specification and initialise the interpreter; otherwise a client's session could be prematurely terminated by another client reinitializing the interpreter in the middle of that session.

The first issue can be resolved by using the client identifier allocated by the API (the field `client`) as a suffix on the name of all objects created by that client. For the current example, there is only one object that falls into this category: the object with the name given by the field `channel`. Thus this can be initialised as follows:

```
channel = "mc" + String.valueOf(client);
```

The second issue can be resolved by checking whether the interpreter has already been initialised before trying to load the model. Both of these issues can therefore be resolved in the class constructor:

```
public Pop3ApiLayer(JTextArea logArea)
{
  try {
    this.logArea = logArea;
    ToolboxClient toolboxClient = new ToolboxClient();
    app = toolboxClient.
          getVDMApplication(new String[]{},
                            ToolType.PP_TOOLBOX);
    client = app.Register();
    channel = "mc" + String.valueOf(client);
    app.PushTag(client);
```

```
      interp = app.GetInterpreter();
      if (!isInitialized())
      {
         loadSpecification();
         typeCheck();
         initInterpreter();
      }
   } catch (Exception e) {
      e.printStackTrace(System.err);
      System.err.println(e.toString());
   }
}
```

This just leaves the issue of how to define isInitialized. This is not supported primitively by the API, in the sense that there is not a method exposed in the API that directly states whether the tool currently contains an initialised model. An approximation used here is to see whether there are currently any parsed files in the tool; if so, it is assumed that another client has already started, so it is not necessary to initialise the model in the interpreter. This gives the following definition:

```
private boolean isInitialized() throws APIError
{
   VDMProject prj = app.GetProject();
   ModuleListHolder moduleList = new ModuleListHolder();
   prj.GetModules(moduleList);
   return moduleList.value.length != 0;
}
```

Exercise 13.6 Under what circumstances is this approximation insufficient? □

Having handled the case of multiple clients in this way, it is now possible to validate the concurrent aspects of the model. For instance, it is possible to check that two clients are not able to access the same mail account at the same time; it is possible to check that changes made on the server by one client (for instance, deletion of a message) persist when the server is accessed by another client.

13.5 Summary

There are many facets to the quality of a model, but the overriding consideration is its fitness for purpose. In this chapter, a range of techniques for evaluation and improving models has been presented. These concentrate on the *internal consistency* of a model and its *external consistency* with client requirements.

Internal consistency checking entails looking for contradictions within the model. Of particular note is the misapplication of partial operators and partial functions or operations and the preservation of invariants in explicit definitions. In implicit definitions, this amounts to checking for the existence of a valid result (satisfiability).

There are many external consistency checking techniques, especially those based on proof and testing, that are considered elsewhere in the literature. In this chapter, the stress has been on the use of an application programmer interface to mediate a model to clients who may not be familiar with the notation in which it is expressed. The approach uses an API layer to mediate between a graphical user interface and the model running in a tool set. The practical details have been illustrated using the VDMTools API and the POP3 example, using the abstraction structure shown in Figure 13.6. In particular this abstraction structure supports straightforward reuse of the GUI and other common portions of code when code generating an implementation. This is discussed further in Chapter 14.

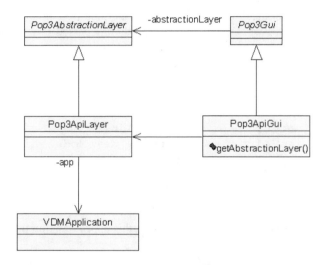

Fig. 13.6: Abstraction structure in Pop3 application

14

Implementing in Java

Although VDM++ is a modelling and specification language, a large subset is suitable for direct generation to executable code. The aim of this chapter is to equip the reader with the ability to recognise models as suitable for code generation and to use VDMTools to generate Java code from VDM++ models.

14.1 Introduction

The emphasis of preceding chapters has been on *models*, particularly their construction, analysis and testing. However, the transition often has to be made from model to implementation, once the developer has gained confidence in the model's qualities. Implementation can be done manually, using developers to handcraft a programming language implementation using the model as a specification. Alternatively, *automatic code generation* can be used to generate an implementation at the press of a button. VDMTools allows automatic generation of an implementation from a VDM++ model in either C++ or Java.

There are advantages and disadvantages to the use of automatic code generation. A full discussion of these issues is beyond the scope of this chapter, but in brief the principal advantage is time saved when the implementation phase of a project is essentially replaced by a single button-press. This comes at the price of a potential loss in the level of abstraction in the model. The nature of the code generated by VDM-Tools is controllable by a number of options within the tool set. For example, one option allows generation of code corresponding to pre- and postconditions in the model. However, the generated code is not intended for human consumption and is therefore not necessarily easy to read.

This chapter will concentrate on automatic code generation with Java as the target language. Java is a modern platform-independent object-oriented programming language, which is well suited to implementation of VDM++ models. The underlying object models of the two languages are similar, allowing a simple transition from VDM++ to Java. Throughout this chapter it is assumed that the Java 2 Standard Edition is used. The specific version can at any stage be found on the book's Web pages.

Code generation into C++ is also supported by VDMTools but this is not described further here.

This chapter is organised as follows: First an overview of the code generation process is presented. This is followed by a more detailed look at the basics of automatic code generation. After this, the complications surrounding concurrency in the underlying VDM++ model are considered. The various strands are drawn together in the penultimate section, in which the model of the POP3 server used in Chapters 12 and 13 is revisited. Finally a summary of the chapter is given. The files in this chapter can be found on the book's Web site.

14.2 Overview of Java Code Generation

The VDMTools automatic Java code generator allows generation of fully compilable Java [Gosling&00] code from a type-correct VDM++ model. This generated code may then be combined with handwritten Java code and a `jar` library distributed with VDMTools to create an executable program. This scheme is illustrated in Figure 14.1, and the main components are considered here.

14.2.1 VDM++ Model

The source VDM++ model must be type-correct so that it is possible to decide which Java constructs should be generated for the various parts of the model. In addition, not all constructs from VDM++ may be code-generated. As a rule of thumb, VDM++ that is not executable in the VDMTools interpreter cannot be the subject of code generation either. A detailed list of the exceptions is given in the VDMTools code generator manual [CGManJavaPP].

14.2.2 Java Files

The generated Java files follow the structure of the corresponding VDM++ classes. Consider the example VDM++ class shown in Figure 14.2. A file `A.java` would be generated from this class by VDMTools. The class A would include a member variable `i` and a method `op`. The class would also include other constructs such as a default constructor and, if generation of concurrency constructs is enabled, code to ensure correct behaviour of classes using synchronisation constraints.

Each generated class member is enclosed within *keep tags*. These are special comments used by the code generator to indicate that changes made in the generated code should be preserved rather than overwritten if the code generator is invoked again on the same class. Keep tags are described in detail in Section 14.3.1.

14.2.3 Handwritten Parts

Before executable code can be obtained, some Java code must be written by hand. At the very least, a main method must be written to act as the entry point to the program;

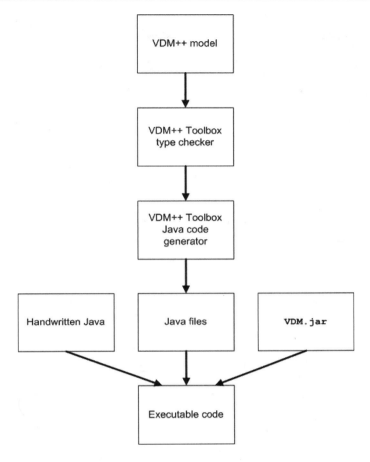

Fig. 14.1: Java code generation of models

at the other extreme the generated code could be a small part of a larger system, the rest of which is handwritten. For example it could be the functional layer lying between a graphical user interface (GUI) and a database. In this case the GUI and the database interaction would be implemented manually.

14.2.4 VDM.jar

A number of utility methods are used by the generated code, for instance, to ensure that equality testing in the generated code is consistent with that defined in VDM++. These utility methods are distributed as a jar file with VDMTools. When compiling and running generated code, this file should be included in the compiler's class path. Documentation for this jar file is distributed with VDMTools.

```
class A

instance variables

i : nat := 0

operations

op : nat ==> ()
op(n) ==
   i := i + n

end A
```

Fig. 14.2: An example VDM++ class for code generation

Having described the basic components involved in Java code generation, the following two sections describe the code generation process in more detail.

14.3 Basic Code Generation

This section describes the basic approach to code generation. Issues relating to concurrency are postponed to the next section. When code is generated from a VDM++ class, the resulting Java class contains members corresponding to constructs in the VDM++. An additional default constructor is generated that takes no arguments and assigns values to the instance variables in the class which, at the VDM++ level, are initialised at the point of declaration.

Table 14.1: Summary of member generation

VDM++ Construct	Java Construct
Instance variable	Class member variable
Value	Static final class member variable
Operation	Class method
Function	Class method

The correspondence between VDM++ class members and their Java counterparts is summarised in Table 14.1. For example, the declaration of the VDM++ instance variable i from Figure 14.2 would lead to the Java declaration of a class member variable:

```
private Integer i = null;
```

Java's class types are used to represent the basic types of VDM++, so here i is a reference to an instance of the Integer class. Note that Java has no direct counterpart to the VDM++ natural number type nat, so, in the generated code, it is not possible to distinguish between values in the model of type nat and int.

In the VDM++ model, the instance variable is initialised at the time of declaration; in the generated code, the variable is initialised in the class's default constructor:

```
public A () throws CGException {
  try {
    i = new Integer(0);
  }
  catch (Exception e){
    e.printStackTrace(System.out);
    System.out.println(e.getMessage());
  }
}
```

All generated methods are declared to throw CGException (defined in VDM.jar); this allows the generated code to handle run-time errors, such as dynamic type errors, gracefully.

The operation op from Figure 14.2 would be generated as a method with the same name:

```
private void op (final Integer n) throws CGException {
  i = UTIL.NumberToInt(
        UTIL.clone(new Integer(i.intValue() +
                                n.intValue())));
}
```

Note here that the method's parameter is declared to be final; this corresponds to the way in which parameters are used in VDM++. The body of the method corresponds to the body of the VDM++ operation; in this example use is made of some methods from the UTIL class (included in VDM.jar).

VDM++ value definitions are translated to static final class member variables in Java. Initialisation of the variable occurs either directly or in a static initialiser, if direct initialisation is not possible. For example, the VDM++ value definition

```
v1 = "literal value";
```

would result in the generation of the following Java

```
private static final String v1 =
                        new String("literal value");
```

whereas the value definition

```
s1 = {1}
```

would yield

```
private static final HashSet s1;
```

and the following static initialiser would be generated:

```
static {
  try {
    HashSet s1temp = new HashSet();
    {
      try {
        HashSet tmpVal_1 = new HashSet();
        tmpVal_1 = new HashSet();
        tmpVal_1.add(new Integer(1));
        s1temp = tmpVal_1;
      } catch (Throwable e) {
        System.out.println(e.getMessage());
      }
    }
    s1 = s1temp;
  } catch (Throwable e){
    e.printStackTrace(System.out);
    System.out.println(e.getMessage());
  }
}
```

Note here that code generated from VDM++ sets uses HashSet from the java.util package. The same package is used for sequences (Vector) and mappings (HashMap).

VDM++ type definitions are treated according to whether the type being defined is a record type. Consider the following type definition:

```
R1 :: a : nat
       b : bool;
```

The code generated from this would use an inner class, as shown in Figure 14.3.

For nonrecord type definitions, if the type being defined is a union of record types, then the resulting Java code is an interface implemented by each inner class corresponding to a record type in the union. For example, if R2 and R3 are defined as

```
R2 :: ;

R3 = R1 | R2;
```

then R3 would be code generated as

```
public static interface R3 {
}
```

and the inner classes R1 and R2 would both implement this interface, as is the case for R1 in Figure 14.3.

All other type definitions are not in fact code generated; instead automatic substitution is performed by the code generator. For example, given the type definition

```
T = nat
```

no code corresponding to this definition would be generated. Wherever T is used in the model, the code generator will generate code as though nat appeared in the same place.

Exercise 14.1 Consider the following class:

```
class A

instance variables
x : nat := 0

operations

public getX : nat ==> nat
getX () ==
   return x;
```

```
public equals : A ==> bool
equals(a) ==
  return a.getX() = x

end A
```

This class can be found on the book's Web site. Create a project containing this class, and automatically generate Java code from it. What is the name of the operation `equals` in the generated code? □

14.3.1 Keep Tags

Automatically generated code is often modified by the developer, for example, to improve its performance or access external resources that were not available at the model level. Such modified code needs to be protected from being overwritten if the rest of the code is to be regenerated from the model. To allow this kind of protection, special comments, known as *keep tags* are generated in the code. A keep tag is generated before and after each top-level method, member variable and inner class in the generated code. The tag before the class member is known as a *start* tag and has the form

```
// ***** VDMTOOLS START Name=<member name> KEEP=NO
```

The tag after the class member is known as an *end* tag and has the form

```
// ***** VDMTOOLS END Name=<member name>
```

For example, the method `op` described earlier would actually be generated with the following keep tags:

```
// ***** VDMTOOLS START Name=op KEEP=NO
  private void op (final Integer n) throws CGException {
    i = UTIL.NumberToInt(
          UTIL.clone(new Integer(i.intValue() +
                                 n.intValue())));
  }
// ***** VDMTOOLS END Name=op
```

By default, any changes to the body of this method will be lost if the class is regenerated. However, if the KEEP=NO part of the start tag was changed to KEEP=YES then any changes to the body of the method would be retained if the class was regenerated.

```
private static class R1 implements R3 , Record {

  public Integer a;
  public Boolean b;

  public R1 () {}

  public R1 (Integer p1, Boolean p2) {
    a = p1;
    b = p2;
  }

  public Object clone () {
    return new R1(a,b);
  }

  public String toString () {
    return "mk_A`R1(" + UTIL.toString(a) +
           "," + UTIL.toString(b) + ")";
  }

  public boolean equals (Object obj) {
    if (!(obj instanceof R1))
      return false;
    else {
      R1 temp = (R1) obj;
      return UTIL.equals(a, temp.a) &&
             UTIL.equals(b, temp.b);
    }
  }

  public int hashCode () {
    return (a == null ? 0 : a.hashCode()) +
           (b == null ? 0 : b.hashCode());
  }
}
```

Fig. 14.3: Code generation of record type

It is vital that the method signature is not changed, because this is assumed by the rest of the generated code.

Keep tags can also be used to add code to a class. The code should be given a name that is unique within the class, and KEEP should be YES. For example, the hashCode method could be overridden:

```
// ***** VDMTOOLS START Name=hashCode KEEP=YES
  public int hashCode () {
    return i.intValue();
  }
// ***** VDMTOOLS END Name=op
```

In fact the name provided in the keep tags need not coincide with any names defined within the keep tags, so, for example, the start tag could have instead been written as Name=myHashCode. Moreover the code within the keep tags may contain more than one class member.

Outside the class definition, keep tags are generated for header comments (with name HeaderComment), a package declaration (with name package) and package imports (with name imports). In this way, user- project- or organisation-specific headers and package names can be used.

Care should be taken when using keep tags, because the relationship between the model and the implementation is weakened by their use. In particular it is easy to forget to modify handwritten code within keep tags when other parts of a model have been updated. For the rest of this chapter, code excerpts are shown without keep tags to improve readability.

Exercise 14.2 Using keep tags, add a method equals to A.java generated in Exercise 14.1, which delegates the equality test to the operation specified in the model. □

14.3.2 Options for Java Code Generation

It is possible to set options in VDMTools to control how code is generated. From the Project menu select Project Options and then the Java code generator tab. This gives the window shown in Figure 14.4. The meaning of each yes/no option (indicated by a checkbox in Figure 14.4) is described here:

Generate only skeletons, except for types: If this option is enabled, operations and functions are generated as normal except that they have dummy bodies. The intention is that the user could then supply an appropriate method body.
Generate only types: This option instructs the code generator to ignore everything but type definitions in the VDM++ model when generating code.
Generate integers as longs: This causes the code generator to use the class Long instead of Integer when generating code for VDM++ nat and int values.

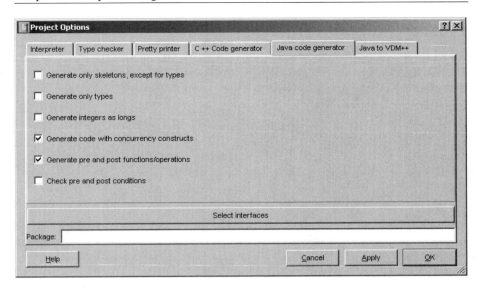

Fig. 14.4: Options for Java code generation

Generate code with concurrency constructs: This option causes the code generator to include special constructs supporting concurrency in the generated code corresponding to concurrency in the model. Without this option, a code generated from a model that uses concurrency will not function correctly. This option is explored further in Section 14.4.

Generate pre and post functions/operations: If this option is enabled, whenever a function or operation pre- or postcondition is encountered, a corresponding method is generated. For example, if a model contains a function f that has a precondition, with this option enabled a method pre_f will be code generated, in addition to f.

Check pre- and postconditions: Enabling this option causes the inclusion of runtime checking of pre- and postconditions. On entry to a method that has a precondition, the precondition is checked; on exit from a method with a postcondition, the postcondition is checked. In either case, failure of the check causes an exception to be thrown.

The Select Interfaces button is described later. The text box labelled Package can be used to give a package name to the generated classes. For example, if this field were set to be dk.vdmbook.examples then the line

package dk.vdmbook.examples;

would appear at the start of each generated file. In addition, the generated files respect the directory structure specified by the package declaration; that is, instead of being generated in the same directory as the project file, the files will be generated in dk/vdmbook/examples with respect to the project file's directory.

Generating Interfaces

Below the list of yes/no options in Figure 14.4 is a button with the text `Select`
`Interfaces`. This allows certain VDM++ classes to be generated as Java interfaces, rather than as classes. To understand this option, consider the VDM++ classes
in Figure 14.5.

```
class D

instance variables

protected i : nat := 0

end D

class E

operations

public op: nat ==> nat
op(n) ==
  is subclass responsibility

end E

class F is subclass of E

end F

class G is subclass of D, F

operations

public op: nat ==> nat
op(n) ==
  return i + n

end G
```

Fig. 14.5: Interface generation example

Normal generation of code from class G is not possible, because it employs multiple class inheritance, which is not supported by Java. However, under special conditions it is possible to generate specific VDM++ classes as Java interfaces rather

than Java classes. The conditions for a class to be generated as a Java interface are: The class may not define any instance variables, nor may it define a thread; any operations or functions defined in the class must have as body `is subclass responsibility` and any superclasses must also be generated as an interface. VDMTools automatically verifies these conditions when the button with text **Select Interfaces** is pressed. With the model shown in Figure 14.5 this gives the dialogue box shown in Figure 14.6.

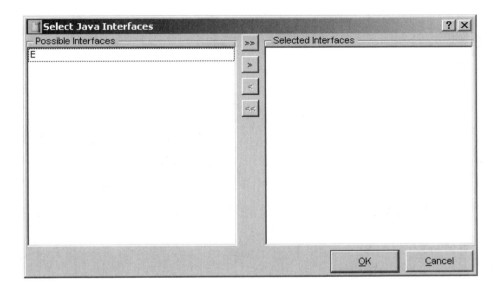

Fig. 14.6: Initial Select Interfaces window

Figure 14.6 shows that initially only one class may be code generated as a Java interface, namely E. Selecting this class and clicking on the button with the symbol ">" informs VDMTools that this class is to be code generated as a Java interface. This causes a recalculation of the possible interfaces, as shown in Figure 14.7.

The class F has now been added to the list of class that could be generated as a Java interface. This is because E, which is the only superclass of F, has now been selected to be generated as an interface, and so the conditions outlined earlier have been fulfilled for F.

Selecting F to be an interface leads to the situation shown in Figure 14.8. Having selected classes E and F to be generated as interfaces, VDMTools Java code generator generates the following code for them:

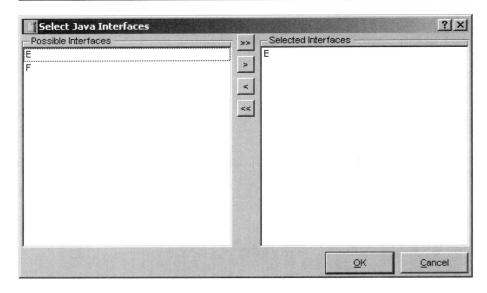

Fig. 14.7: Select Interfaces window after selection of E

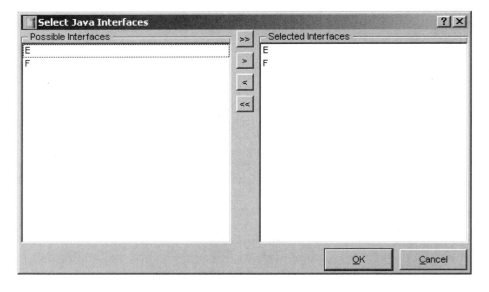

Fig. 14.8: Select Interfaces window after selection of E and F

```
public interface E {
  abstract public Integer op (final Integer n)
  throws CGException ;
}

public interface F extends E {
}
```

It is then possible for the multiple inheritance used by class G to be resolved. The Java class generated from G is as follows:

```
public class G extends D implements F {

  public G () throws CGException {}

  public Integer op (final Integer n)
                   throws CGException {
    return new Integer(i.intValue() + n.intValue());
  }
}
```

Note that if G had not contained an implementation of op, it would have been generated as an abstract class.

Exercise 14.3 Consider the following classes:

```
class B

end B

class C is subclass of B

end C

class D is subclass of B

end D

class E is subclass of D

end E

class F is subclass of C, D
```

```
end F
```

Code-generate all of these classes to valid Java classes. □

14.4 Code Generation and Concurrency

The VDMTools Java code generator can generate code supporting the concurrency constructs of VDM++ if the option **Generate code with concurrency constructs** is selected. When this option is enabled, extra code is generated in the form of additional class members and extra code in method bodies. All instance variables are generated as `volatile` class members. When this option is not enabled, this extra code is not generated, but concurrent behaviour in the model will not be emulated in the generated code. By default this option is enabled.

In this section the extra constructs generated for concurrency are described. First, code generation from permission predicates is described, and thereafter the treatment of threads is presented.

14.4.1 Permission Predicates

Several changes to the basic translation process are made when code is generated to represent the behaviour modelled by permission predicates:

- The generated class implements the interface `EvaluatePP`; an interface defined in `VDM.jar`.
- An inner *sentinel* class is generated. This defines symbolic constants for the operations defined in the class.
- A member variable called `sentinel` and a method called `setSentinel` are generated.
- A method called `evaluatePP` is generated, representing the permission predicates defined in the corresponding VDM++ class.
- Each generated method corresponding to an operation in the model has extra code at the beginning and end.

As an example, consider the VDM++ class G:

```
class G

instance variables

i : int := 0

operations
```

```
public put : nat ==> ()
put(n) ==
  i := n;

public get : () ==> nat
get() ==
  return i;

sync
  mutex(put,get)

end G
```

In the generated Java class, the class header is:

```
public class G implements EvaluatePP
```

The inner class GSentinel is generated as a subclass of Sentinel (defined in VDM.jar). Sentinel defines member variables and methods that emulate the behaviour of VDM++ history counters. GSentinel then provides a specialization of Sentinel for G:

```
class GSentinel extends Sentinel
{
  public final int get = 0;
  public final int put = 1;
  public final int nr_functions = 2;

  public GSentinel () throws CGException {}

  public GSentinel (EvaluatePP instance)
  throws CGException
  {
    init(nr_functions, instance);
  }
}
```

A member variable sentinel is declared to be an instance of Sentinel:

```
volatile Sentinel sentinel;
```

and the method `setSentinel` initialises this member variable:

```
public void setSentinel ()
{
  try {
    sentinel = new GSentinel(this);
  } catch (CGException e) {
    System.out.println(e.getMessage());
  }
}
```

A method `evaluatePP` is defined, which represents evaluation of a permission predicate. All permission predicates are converted into expressions on history counters; then the sentinel is used to extract these history counters:

```
public Boolean evaluatePP (int fnr) throws CGException {
  Boolean temp;
  switch(fnr) {
  case 0: {
    temp = new Boolean(UTIL.equals(
             new Integer(sentinel.active[
                ((GSentinel) sentinel).get] +
          sentinel.active[((GSentinel) sentinel).put]),
                         new Integer(0)));
    return temp;
  }
  case 1: {
    temp = new Boolean(UTIL.equals(
             new Integer(
               sentinel.active[((GSentinel) sentinel).get]
               +
               sentinel.active[((GSentinel) sentinel).put]),
               new Integer(0)));
    return temp;
  }
  }
  return new Boolean(true);
}
```

This method is called with the symbolic constant defined in `GSentinel` for the method whose permission predicate is being evaluated.

To ensure that history counters are correctly incremented, extra code is added at the start and end of methods corresponding to operations. For instance, the operation put is code-generated as:

```
public void put (final Integer n) throws CGException
{
  sentinel.entering(((GSentinel) sentinel).put);
  try {
    i = UTIL.NumberToInt(UTIL.clone(n));
  } finally {
    sentinel.leaving(((GSentinel) sentinel).put);
  }
}
```

On entering the method, sentinel.entering is called. This method call does three things: first it increments the request counter for this method of this object; then it evaluates the permission predicate for this method. If this evaluates to true, control returns to the calling environment (the put method). If this evaluates to false, the method blocks, and another thread is activated (or if no other thread can be activated, a deadlock has been reached).

When the method call is complete, the finally clause uses the call to the method sentinel.leaving to increment the finished history counter.

14.4.2 Threads

VDM++ classes that include a thread declaration are translated to Java classes that take advantage of Java's own thread constructs. To this end such classes implement the Runnable interface and provide a run method that is invoked when the thread is started. Thus the run method corresponds to the thread body specified in the VDM++ class. Further, a member variable procThread is generated, which is an instance of VDMThread, a simple wrapper class whose objective is to allow support for VDM++'s threadid keyword. Finally, a start method is generated, which initialises procThread and then starts the thread. The following example demonstrates this. Consider the VDM++ class H defined here:

```
class H

thread

( dcl count : nat := 0,
      g : G := new G();
  while count < 10 do
  ( count := g.get();
    g.put(count+1)
```

```
   )
 )

 end H
```

The generated class has header

```
public class H implements EvaluatePP , Runnable
```

The thread body is generated as the method run:

```
  public void run () {
    try {
      Integer count = new Integer(0);
      G g = new G();
      while ( new Boolean((count.intValue()) <
        (new Integer(10).intValue()))).booleanValue()){
        Integer rhs_6 = null;
        rhs_6 = g.get();
        count = UTIL.NumberToInt(UTIL.clone(rhs_6));
        g.put(new Integer(count.intValue() +
                        new Integer(1).intValue()));
      }
    } catch (Throwable e) {
      System.out.println(e.getMessage());
    }
  }
```

The member variable procThread is declared:

```
VDMThread procThread;
```

This variable is initialised and used in the method start:

```
public void start () throws CGException {
  procThread = new VDMThread(this);
  procThread.start();
}
```

Generation of a call to this method corresponds to the use of a start statement with instances of this class in the model.

Exercise 14.4 Recall the `Buffer` class from Chapter 12. This can be found on the book's Web site. Generate Java code from this with the concurrency option turned both on and off. Compare the generated files. □

14.5 Code Generation of POP3

This section pulls together the idea strands presented in the preceding sections, using a more realistic example based on the POP3 server from Chapter 12. In particular, the GUI used for external consistency assessment of the VDM++ model using VDM-Tools API, as described in Chapter 13, is reused with the generated code. This is a particularly useful design pattern when both assessment of external consistency and code generation are performed, and it is therefore discussed in some detail in Section 14.5.2. First, the structure of the POP3 server is described, and the relationship between handwritten and generated code is considered.

14.5.1 Overview of Server Side

The POP3 model presented in Chapter 12 models a server, with an abstraction of a network socket that clients use to connect to the server. In Chapter 13 a graphical client was used to interact with the server, with communication between the client and the server essentially via CORBA. In the case of the code-generated server, some code must be handwritten to translate between the actual socket used by client and server, and the abstractions used in the model. Two classes are handwritten for this reason: The class `POP3ServerInterface` listens on a socket for clients trying to initiate contact with the server; the class `InputMessageChannel` is a wrapper class for `MessageChannel` from the model, using input streams to send and receive messages between the client and the server. The class diagram in Figure 14.9 gives an overview of this structure. In this figure, the classes in white are handwritten, and

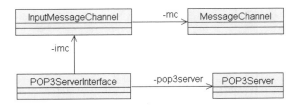

Fig. 14.9: Structure of code-generated POP3 server

those in grey are code-generated; only those generated classes that interface with the handwritten ones are shown.

The POP3 model uses concurrency to allow multiple clients to access mailboxes simultaneously. It is therefore necessary to set the option for concurrent code generation in VDMTools. With this set, the code automatically generated from the model's permission predicates ensures that communication between the clients and server in the generated code proceeds in the same way as it did in the model.

The Class `POP3ServerInterface`

This class acts as the public interface to the POP3 server; it has a thread that listens for clients wanting to initiate a POP3 session. The class has three principal member variables:

```
ServerSocket serverSocket;
POP3Server pop3server;
MessageChannelBuffer ch;
```

`ServerSocket` (from the package `java.net`) allows the server to listen to a particular port; `POP3Server` and `MessageChannelBuffer` are code-generated from the VDM++ model. The main functionality in the `POP3ServerInterface` class is in its `run` method. This method starts the `pop3server` thread then waits for contact from clients. When a client makes contact, an `InputMessageChannel` object is created, and the client then uses this to communicate with the server. In the meantime the server returns to listening for clients. This method is shown in Figure 14.10.

The Class `InputMessageChannel`

The `InputMessageChannel` class wraps a `MessageChannel` object. It has a thread that mediates between this `MessageChannel` and the client, which it uses object streams to communicate with. It has three member variables:

```
ObjectInputStream inputStream;
ObjectOutputStream outputStream;
MessageChannel mc = null;
```

The constructor takes the socket that the client used to initiate contact with the server and initialises the object streams:

```
public InputMessageChannel(Socket socket)
throws IOException
{
  outputStream = new ObjectOutputStream(
                        socket.getOutputStream());
```

```java
public void run(){
  try {
    pop3server.start();
    serverSocket = new ServerSocket(4444);
  } catch (IOException e) {
    System.out.println(
      "Could not listen on port 4444");
    System.exit(-1);
  } catch (CGException e){
    System.out.println(e.toString());
    System.exit(-1);
  }

  while (true) {
    Socket clientSocket = null;
    try {
      clientSocket = serverSocket.accept();
      InputMessageChannel imc =
        new InputMessageChannel(clientSocket);
      ch.Put(imc.getMessageChannel());
      imc.start();
    } catch (IOException e){
      System.out.println(
        "Accept failed on port 4444");
      System.exit(-1);
    } catch (CGException e){
      System.out.println(e.toString());
    }
  }
}
```

Fig. 14.10: run method from POP3ServerInterface

```java
inputStream = new ObjectInputStream(
                      socket.getInputStream());
try {
    mc = new MessageChannel();
} catch (CGException e){
    System.out.println(e.toString());
}
}
```

The class's thread, defined by its `run` method, repeatedly receives messages from the client, which it forwards to the `MessageChannel`. It then sends the response from the `MessageChannel` back to the client:

```java
public void run() {
  while (true) {
    try {
      Object command = inputStream.readObject();
      mc.ClientSend(
        (POP3Types.ClientCommand) command);
      Object response = mc.ClientListen();
        outputStream.writeObject(response);
    } catch (Exception e) {
      System.out.println(e.toString());
    }
  }
}
```

14.5.2 Overview of Client Side

To interact with the POP3 server, a POP3 client is required. Because the model presented in Chapter 12 is not a complete implementation of the POP3 protocol, using a standard POP3 client could be problematic; however in Chapter 13 a POP3 client was implemented specifically for use with the POP3 server model of Chapter 12. It would be desirable if as much of this client as possible could be reused. The class structure for the Java code in this chapter and Chapter 12 explicitly supports extensive reuse. An overview of this structure is shown in Figure 14.11.

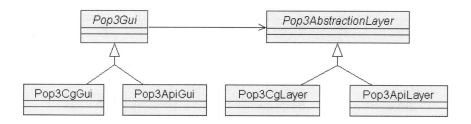

Fig. 14.11: Structure of code-generated POP3 server

The basic functionality of the POP3 client is captured by the abstract class called `Pop3Gui`. This class builds the interface and then connects the various actions in the interface to functionality defined by `Pop3AbstractionLayer`. `Pop3Gui` is abstract because it defines an abstract method `getAbstractionLayer()`, which returns a reference to the abstraction layer object; in the case of the subclass `Pop3CgGui`, `getAbstractionLayer()` returns a reference to a `Pop3CgLayer` object, whereas for `Pop3ApiGui` a reference to a `Pop3ApiLayer` object is returned. The cornerstone of the reuse is `Pop3AbstractionLayer`: This enables the client GUI to be developed independently of how the functionality is defined. Concrete subclasses of `Pop3AbstractionLayer` then define the functionality, which could be in terms of interaction with VDMTools API, automatic code generation from VDMTools, or even a handwritten implementation of the model. `Pop3-AbstractionLayer` defines abstract methods for all of the functionality required by the GUI. An overview of the class is shown in Figure 14.12. For each interaction

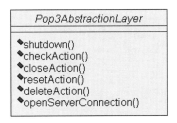

Fig. 14.12: `Pop3AbstractionLayer` class

between the client GUI and server there is an abstract method in `Pop3Abstrac-tionLayer`. Concrete subclasses can implement these methods as they see fit. As an example, two different implementations of `openServerConnection` are shown in Figures 14.13 and 14.14.

In fact, this is an example of a more general pattern that can be used whenever a GUI is developed to assess the quality of a model using VDMTools API, followed by code generation of the model. Figure 14.15 illustrates the generic structure of this pattern.

14.6 Summary

This chapter has focused on the implementation of a model in Java. The issues regarding automatic Java code generation in VDMTools have been discussed, and the different options that can be used to control the way in which code is generated have also been described. Finally, practical issues that arise when a realistic model is code generated were described, and a general pattern for code reuse between code generation

```
public boolean openServerConnection(String username,
                                    String password,
                                    StringBuffer response)
{
  try {
    EvalCmd("create " + channel +
              " := new MessageChannel()");
    EvalCmd("debug ch.Put(" + channel + ")");
    boolean status = executeCommand(
                       "USER",
                       new String[]{"\""+username+"\""},
                       response);
    if (!status)
      return false;
    status = executeCommand(
                       "PASS",
                       new String[]{"\""+password+"\""},
                       response);
    return status;
  } catch (Exception e) {
    e.printStackTrace(System.err);
    return false;
  }
}
```

Fig. 14.13: `Pop3ApiLayer.openServerConnection`

and quality assessment of a model was presented. Project-specific issues such as integrating VDMTools code generation into version control systems were not discussed, but experience suggests that such issues can be resolved without great difficulty.

```java
public boolean openServerConnection(String username,
                                    String password,
                                    StringBuffer messageBuffer)
{
  try {
    String hostname = System.getProperty("POP3HOST",
                                          "localhost");
    Socket serverSocket = new Socket(hostname, 4444);
    outputStream = new ObjectOutputStream(
                        serverSocket.getOutputStream());
    inputStream = new ObjectInputStream(
                        serverSocket.getInputStream());
    Object responseObj = sendCommandResponse(
                        new POP3Types.USER(username));
    if (checkResponse(responseObj, messageBuffer)) {
      responseObj = sendCommandResponse(
                        new POP3Types.PASS(password));
      return checkResponse(responseObj,
                        messageBuffer);
    }
    else
      return false;
  } catch (Exception e){
    messageBuffer.append(e.toString());
    return false;
  }
}
```

Fig. 14.14: `Pop3CgLayer.openServerConnection`

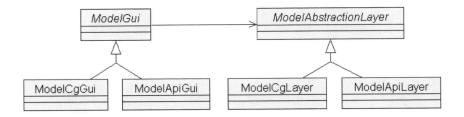

Fig. 14.15: Class structure for abstraction layer pattern

Appendices

A

Solutions to Exercises

Solutions for most of the exercises are given in this appendix. However, for exercises where the solutions are large and can probably only be feasibly handled using the VDMTools the solutions are only given on the book's Web pages.

Solutions for Chapter 4

Exercise 4.1

All of the concepts are animals, so it seems reasonable to provide an overall superclass called `Animal`. A common solution then would be to make birds and fish subclasses of `Animal` and to treat the individual species as the next layer of subclasses. The corresponding UML class diagram appears in Figure A.1. In VDM++, this would look like:

```
class Animal

end Animal
```

```
class Bird is subclass of Animal

end Bird
```

```
class Fish is subclass of Animal

end Fish
```

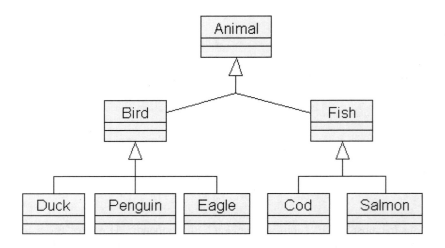

Fig. A.1: Animal and its subclasses

```
class Duck is subclass of Bird

end Duck
```

```
class Penguin is subclass of Bird

end Penguin
```

```
class Eagle is subclass of Bird

end Eagle
```

```
class Cod is subclass of Fish

end Cod
```

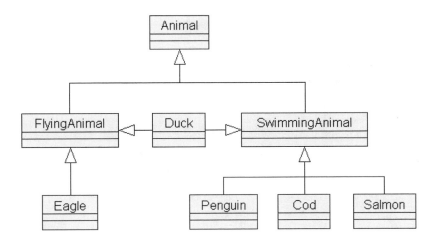

Fig. A.2: Animals with multiple inheritance

```
class Salmon is subclass of Fish

end Salmon
```

Exercise 4.2

To take the functionality into account, the ability to swim or fly could form the basis of the class structure. The subclasses of animal would then be flying and swimming animals. The species inherit the ability to swim or fly, or do both, from these classes. This is a cleaner use of inheritance for the modelling problem at hand. It is best if the superclasses satisfy the properties that the subclasses satisfy. The UML class diagram in shown in Figure A.2 and the VDM++ would be as follows:

```
class Animal

end Animal
```

```
class FlyingAnimal is subclass of Animal

end FlyingAnimal
```

```
class SwimmingAnimal is subclass of Animal

end SwimmingAnimal
```

```
class Duck is subclass of FlyingAnimal, SwimmingAnimal

end Duck
```

```
class Penguin is subclass of SwimmingAnimal

end Penguin
```

```
class Eagle is subclass of FlyingAnimal

end Eagle
```

```
class Cod is subclass of SwimmingAnimal

end Cod
```

```
class Salmon is subclass of SwimmingAnimal

end Salmon
```

This is clearly a cleaner use of inheritance. In general it is best if the superclasses satisfy all the properties that subclasses satisfy.

Exercise 4.3

```
class NonSquare

 instance variables
   width  : nat;
   height : nat;
   inv width <> height

end NonSquare
```

Exercise 4.4

First, it should be noted that having a door is not specific to shops; *all* units have doors. Therefore the instance variable that represents the door should be part of class Unit, not of class Shop. The invariant over the instance variables is that if the shop is closed *and* the door is open *then* the alarm should be ringing. In VDM++ this can be modelled as follows:

```
class Unit

instance variables
openDoor: bool := false

. . .

end Unit

class Shop is subclass of Unit

instance variables
isOpen       : bool := false;
alarmRinging : bool := false;
inv (openEntrance and not isOpen) => alarmRinging

. . .

end Shop
```

Exercise 4.5

This could be expressed as an enumerated type using the combination of VDM++
quote types and a union type:

```
<Open> | <Closed>
```

Exercise 4.6

In VDM++ this can be written as follows inside a class, for example, called `Structure`:

```
class Structure

 types

 Tree = nat | Node;

 Node :: value : nat
         left  : Tree
         right : Tree

end Structure
```

Exercise 4.7

By making the tree type optional, it is possible to have nil subtrees:

```
class Structure

types

 Tree = [nat] | Node;

 Node :: value : nat
         left  : Tree
         right : Tree

end Structure
```

Exercise 4.8

Record types can be used for emulating strong typing in VDM++. By defining two types

```
A :: nat
```

and

```
B :: nat
```

two type *definitions* A and B are provided, where the values cannot be compared because the *tags* A and B are different.

Solutions for Chapter 5

Exercise 5.1

```
Sample(pop: set of CPRNo, sexreqd:Patient'Sex, size:nat1)
       s:set of CPRNo
pre card pop >= size and
    forall p in set pop & Sex(p) = sexreqd
post card s = size and
     forall p in set s &
        p in set pop and Sex(p) = sexreqd
```

Note that the requirement on the sex of the members of the sample can be removed from the postcondition, because it is guaranteed by the precondition:

```
Sample(pop: set of CPRNo, sexreqd:Patient'Sex, size:nat1)
       s:set of CPRNo
pre card pop >= size and
    forall p in set pop & Sex(p) = sexreqd
post card s = size and
     forall p in set s & p in set pop
```

Finally, let nil stand for a suitable error value:

Validated Designs for Object-oriented Systems

```
Sample(pop: set of CPRNo, sexreqd:Patient'Sex, size:nat1)
       s:[set of CPRNo]
post if (card pop >= size and
            forall p in set pop & Sex(p) = sexreqd)
       then (card s = size and
              forall p in set s & p in set pop)
       else s=nil
```

Exercise 5.2

1. 28.5
2. 28.5
3. <Even>
4. "Unknown"
5. mk_Coord(5,0)

Exercise 5.3

```
fac: nat ==> nat
fac(n) ==
(dcl result: nat := 1,
     index : nat := n;
 while index >= 1 do
 ( result := result * index;
   index := index - 1
 );
 return result;
);
```

Solutions for Chapter 6

Exercise 6.1

1. equal
2. not equal
3. not equal

Exercise 6.2

1. true
2. false

Exercise 6.3

1. true
2. false

Exercise 6.4

1. {4, 8}
2. {{10, 11, 12}, {11, 12, 13}, {12, 13, 14}}
3. {0, 4}

Exercise 6.5

1. false
2. true
3. false

Exercise 6.6

```
class Route
...
functions
static public TakeStep: set of Point -> set of Point
TakeStep(pts) ==
  let laterPoints = {pt | pt in set pts
                       & pt.GetIndex() <> 1}
  in
    {p.TakeStep() | p in set laterPoints};
end Route
```

Exercise 6.7

1. {<Apple>, <Orange>, <Banana>, <Apple>, <Grape>}
2. {1, 2, 3, 4, 5, 6, 7, 8, 9, 10}
3. {{1, 2, 3}, {4, 5}, {7, -1}, {}}

Exercise 6.8

1. {<Apple>, <Orange>, <Grape>, <Cherry>, <Pineapple>}
2. {}
3. {{}}

Exercise 6.9

```
Dunion: set of set of nat -> set of nat
Dunion(ss) ==
  if ss = {}
  then {}
  else let s in set ss
        in s union Dunion ( ss \ {s})
```

Exercise 6.10

```
class Route
...
operations
GetNext: Point ==> Point
GetNext(pt) ==
  return GetAtIndexFromIndex(pt.GetIndex() + 1);
GetAtIndexFromIndex: nat ==> Point
GetAtIndexFromIndex(index) ==
  return GetPointAtIndex(points, index);
end Route
```

Exercise 6.11

```
class Route
...
operations
GetFirst: () ==> Point
GetFirst() ==
  return GetPointAtIndex(points, 1);
GetLast: () ==> Point
```

```
GetLast() ==
  return GetPointAtIndex(points, card points);
end Route
```

Exercise 6.12

1. {<Apple>}
2. {}
3. {}

Exercise 6.13

1. {<Orange>}
2. {}
3. {}

Exercise 6.14

```
Dinter: set of set of nat -> set of nat
Dinter(ss) ==
  let s in set ss
  in
    if card ss = 1
    then s
    else s inter Dinter (ss \ {s})
pre ss <> {}
```

Exercise 6.15

```
public static AvoidanceRoutes(
                obstacles: set of (nat * nat),
                currentPosition: Point,
                nextWaypoint: Point)
                routes: set of Route
post (forall r in set routes &
        r.GetFirst().GetCoord() =
          currentPosition.GetCoord() and
        r.GetLast().GetCoord() =
```

```
        nextWaypoint.GetCoord() and
      r.GetCoords() inter obstacles = {}
  ) and
  ( exists r: Route &
      r.GetFirst().GetCoord() =
      currentPosition.GetCoord() and
      r.GetLast().GetCoord() =
      nextWaypoint.GetCoord() and
      r.GetCoords() inter obstacles = {}
  ) => routes <> {}
```

Exercise 6.16

1. false
2. true
3. true

Exercise 6.17

```
Subset: set of nat * set of nat -> bool
Subset(s1, s2) ==
  forall x in set s1 &
    x in set s2
```

Exercise 6.18

1. false
2. true
3. false

Exercise 6.19

1. {{}, {1}, {2}, {4}, {1,2}, {1,4}, {2,4}, {1, 2, 4}}
2. {{}, {<Banana>}}
3. {{}}

Exercise 6.20

```
Power: set of nat -> set of set of nat
Power(s) ==
  if s = {}
  then {{}}
  else let x in set s
       in
          let rest = Power (s \ {x})
          in
             rest union
             {r union {x} | r in set rest}
```

Solutions for Chapter 7

Exercise 7.1

1. not equal
2. equal
3. not equal
4. equal

Exercise 7.2

1. []
2. []
3. [{1}, {4, 16}]
4. [[1], [2], [3], [4], [5]]

Exercise 7.3

```
[6 - i | i in set {1, 2, 3, 4, 5}]
```

Exercise 7.4

1. 4
2. 1
3. 0
4. 1

Exercise 7.5

1. true
2. [true, false]
3. []
4. [[false, true]]

Exercise 7.6

```
FilterSmall: seq of int * int -> seq of int
FilterSmall(s, n) ==
  [s(i) | i in set inds s & n >= s(i)]
```

Exercise 7.7

1. [5, 3, 3, 9]
2. [9, 3, 2, 3]
3. [9]
4. []

Exercise 7.8

1. []
2. [9, 3, 2, 3]
3. [true, false, [true]]
4. [[]]

Exercise 7.9

```
for i in set inds passInAccount do
  accSpeed := accSpeed + passInAccount(i)
```

Exercise 7.10

```
class PassageSensor is subclass of Sensor

functions
SeqSum: nat1 * seq of nat -> nat
SeqSum (n, s) ==
  if n = 1
  then hd s
  else hd s + SeqSum(n-1, tl s)

...

operations
public AverageSpeed: nat1 ==> CWS'Speed
AverageSpeed(numberOfPassages) ==
let passInAccount = passages(1,..., numberOfPassages) in
    return(SeqSum (numberOfPassages, passInAccount) /
          numberOfPassages)
pre len passages >= numberOfPassages

...

end PassageSensor
```

Exercise 7.11

```
public ShowSignal: CWS'Location *
                   CongestionMonitor'Signal ==> ()
ShowSignal(location, signal) ==
(if location <> len as then
 let downstream = as(location + 1),
     actuator   = as(location)
 in
   -- Set the right signal at the location itself
   ShowSignalAtLoc(signal,downstream,actuator);

 if location <> 1 then
 let upstream   = as(location - 1) in
   -- Set the right signal upstream
   ShowSignalUpstream(signal,upstream)
```

```
)
pre location in set inds as and
    (signal = <NoWarning> or
     signal = <CongestionWarning>);
```

Exercise 7.12

1. [true, false, 4]
2. [4, 4, 2, 2]

Exercise 7.13

```
ConvertNum2String: nat1 -> seq of char
ConvertNum2String (n) ==
let lastdigit = cases n mod 10:
                  0 -> "0",
                  1 -> "1",
                  2 -> "2",
                  3 -> "3",
                  4 -> "4",
                  5 -> "5",
                  6 -> "6",
                  7 -> "7",
                  8 -> "8",
                  9 -> "9"
                end
in if n < 10
then lastdigit
else ConvertNum2String (n div 10) ^ lastdigit
```

Exercise 7.14

```
inv len messagelog = len locations and
    forall i in set inds messagelog &
    let locstring = ConverNum2String(locations(i)) in
       locstring = messagelog(i)(len messagelog(i) -
                                 len locstring, ...,
                                 len messagelog (i))
```

Exercise 7.15

```
WriteLog (message: seq1 of char, location: CWS'Location)
ext wr messagelog: seq of seq1 of char
    wr locations : seq of CWS'Location
post messagelog = ~messagelog ^
                    [message ^ ConvertNum2String(location)]
        and
        locations  = ~locations ^ [location]
```

Exercise 7.16

1. {<Red>}
2. {2, 4, 5}
3. {[]}
4. {[5]}

Exercise 7.17

The modified model can be found on the book's Web pages.

Solutions for Chapter 8

Exercise 8.1

1. not equal
2. equal

Exercise 8.2

1. {1 |-> 0.5, 4 |-> 1, 16 |-> 2, 25 |-> 2.5}
2. {1 |-> 0.5}

Exercise 8.3

1. {}
2. {3, 4}
3. {<Apple>}
4. {3, 4}

Exercise 8.4

The operation `AddCongestionMonitor` can be written as:

```
class CWS
...
public AddCongestionMonitor: Location ==> ()
AddCongestionMonitor(loc) ==
let newmon = new CongestionMonitor(loc, sensors (loc),
                                    ns, op)
in
  roadNetwork := roadNetwork munion { loc |-> newmon }
pre loc not in set dom roadNetwork and
    loc in set dom sensors
end CWS
```

Exercise 8.5

```
inv forall locationset in set rng am &
       not exists loc in set locationset &
          exists otherloc in set rng am locationset &
             loc = otherloc
```

Exercise 8.6

1. `{1 |-> 5, 3 |-> 1, 5 |-> 1}`
2. `{1 |-> 5, 3 |-> 1, 5 |-> 1}`
3. This is not allowed; the value of this expression is `undefined`.
4. `{1 |-> 5, 3 |-> 1, 5 |-> 1}`

Exercise 8.7

Because otherwise there would not be a unique mapping from every domain element and thus there would be two mappings with the same domain value but different range values. This is not allowed for VDM++ mappings.

Exercise 8.8

The operation `GetLocations` can be written as:

```
class NameServer
...
public GetLocations: () ==> set of CWS'Location
GetLocations() ==
  return dunion rng am
end NameServer
```

Exercise 8.9

That depends on the actual algorithm being used to calculate the congestion status:
If that algorithm is dependent on the lane position of the passage sensors, then a set
cannot be used as it does not include that information.

Exercise 8.10

```
public AddActuator: CWS'Location ==> ()
AddActuator(loc) ==
  def actuator = new Actuator()
  in
  ( as := as munion {loc |-> actuator};
    ns.SetLocation(self, loc)
  )
pre loc not in set dom as;
```

Exercise 8.11

1. {<a> |-> 1, <c> |-> 3}
2. {<d> |-> 1}
3. { |-> }
4. {<a> |-> 1, <c> |-> 3, <d> |-> 1}

Exercise 8.12

The adjusted operation in class `ActuatorManager` becomes:

```
public RemoveActuator: Actuator ==> ()
RemoveActuator(actuator) ==
  (as := as :-> {actuator};
   ns.RemoveLocation(inverse as (actuator))
  )
```

and the corresponding operation in class `NameServer` can be written as:

```
public RemoveLocation: Location ==> ()
RemoveLocation(location) ==
  let manager in set dom am
  be st location in set am(manager)
  in
      am := am ++ {manager |-> am(manager) \ {location}}
```

Exercise 8.13

The operation `ReplaceActuatorManager` can be rewritten using the range-restrict-to operator as:

```
public ReplaceActuator: Actuator * Actuator ==> ()
ReplaceActuator (actuator, newActuator) ==
let oldActuator = as :> {actuator}
in
  let location in set dom oldActuator
  in
    as := as ++ {location |-> newActuator}
pre actuator in set rng as;
```

Using a let-be-such-that-statement the operation can be rewritten as:

```
public ReplaceActuator: Actuator * Actuator ==> ()
ReplaceActuator (actuator, newActuator) ==
let location in set dom newActuator be st
    newActuator(location) = {actuator}
in
  as := as ++ {location |-> newActuator}
pre actuator in set rng as;
```

Solutions for Chapter 9

Exercise 9.1

The easiest way to increase the cipher strength of a monoalphabetic cipher is to use some arbitrary random order of symbols instead of "A".."Z". This is only a very weak increase because the principal problem of monoalphabetic ciphers is the fact that the symbol frequency (the number of times a symbol occurs on average in a language) is left untouched. If a suitable number of encrypted messages are intercepted and analysed statistically, it is not hard to guess the encoding distance by comparing the symbol frequencies of the encrypted text against the symbol frequencies of the plain text language.

Exercise 9.2

```
class Component

operations

public Successors: () ==> set of Component
Successors () ==
   (dcl res : set of Component := {self},
        tmp : [Component]        := next;
    while tmp <> nil do
      (res := res union {tmp};
       tmp := tmp.next );
    return res
   )

...

end Component
```

Exercise 9.3

The instance variable `next` is directly accessible to any subclass of class `Component`, due to the `protected` access qualifier. In principle, any subclass can modify `next` *without* using the `SetNext` operation, so it is not hard to create circular structures. The solution is to either add a class invariant or change the access qualifier for the instance variable `next` to `private`, so that it is only possible to modify the instance variable through the operation `SetNext`, hence enforcing the noncircularity requirement. In the latter case it is also needed to add an operation `protected GetNext` to allow access again to the value stored in the instance variable `next` from all sub-classes of `Component`.

Exercise 9.4

See Table A.1.

Table A.1: Solution Encode and Decode behaviour

Configuration	Encode	Decode
Plugboard	Encode(1) = 3	Decode(1) = 3
	Encode(2) = 2	Decode(2) = 2
	Encode(3) = 1	Decode(3) = 1
	Encode(4) = 4	Decode(4) = 4
Rotor	Encode(1) = 3	Decode(1) = 2
	Encode(2) = 1	Decode(2) = 3
	Encode(3) = 2	Decode(3) = 1
	Encode(4) = 4	Decode(4) = 4
Reflector	Encode(1) = 3	Decode(1) = 3
	Encode(2) = 4	Decode(2) = 4
	Encode(3) = 1	Decode(3) = 1
	Encode(4) = 2	Decode(4) = 2

Exercise 9.5

{1 |-> 1, 2 |-> 3, 3 |-> 6, 4 |-> 2, 5 |-> 5, 6 |-> 4}

Exercise 9.6

The solution is similar to the strategy used in class `Component`. First, the class `Test` needs to be extended:

```
class Test

operations
  public Contains : Test ==> bool
  Contains (tst) == return self = tst

...

end Test
```

It is now possible to express the "strict hierarchy" requirement inside the class `TestSuite` by redefining the operation `Contains` and adding a precondition to the operation `AddTest`:

```
class TestSuite

operations

public Contains : Test ==> bool
Contains (tst) ==
  return Test'Contains(tst) or
    exists i in set inds tests &
      tests(i).Contains(tst);

public AddTest: Test ==> ()
AddTest(test) ==
  tests := tests ^ [test]
pre not Contains(test)

...

end TestSuite
```

Solutions for Chapter 10

Exercise 10.1

```
class A

instance variables
  private theC : C;
  private attribA : nat;

end A
```

```
class B

instance variables
  protected theA : A;
```

```
  protected attribB : bool;

end B
```

```
class C

instance variables
  public theB : B;
  public attribC : char;

end C
```

Exercise 10.2

See Figure A.3.

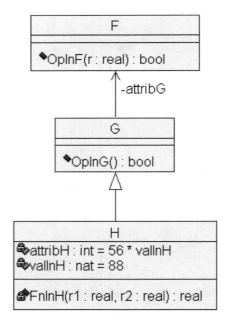

Fig. A.3: Class diagram for Exercise 10.2

Exercise 10.3

```
class C

instance variables
  public theB : B;
  public attribC : char;

end C
```

```
class D is subclass of C

operations
  protected OpInD : set of nat ==> nat
  OpInD(s) ==
    is not yet specified;

end D
```

```
class E is subclass of C

instance variables
  private attribE : real;

end E
```

```
class F is subclass of D, E

operations
  public OpInF : real ==> bool
  OpInF(r) ==
    is not yet specified;

end F
```

Exercise 10.4

```
class J

instance variables

theorderedI : seq of I;
attribJ     : set of real := {};

end J
```

```
class I

instance variables

theorderedA : seq of A;
attribI     : seq of bool;

end I
```

```
class A

instance variables

private attribA : nat;

end A
```

Exercise 10.5

See Figure A.4.

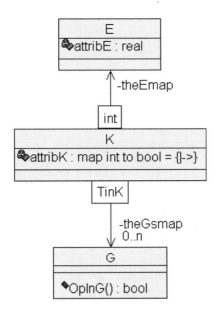

Fig. A.4: Class diagram for Exercise 10.5

Exercise 10.6

```
class D

operations

protected
OpInD : set of nat ==> nat
OpInD(s) ==
   is not yet specified;

end D
```

```
class E

instance variables
```

```
private attribE : real;

end E
```

Exercise 10.7

```
class M

instance variables

theRs : set of R;
protected attribM : bool;

end M
```

```
class N is subclass of M

instance variables

public theO : O;

operations

public
OpInN : set of int ==> nat
OpInN(s) ==
  is not yet specified;

end N
```

```
class O

instance variables

public attribO : set of nat;

end O
```

```
class P is subclass of N, O

instance variables

public theQmapping : map int to Q;

functions

public
FnInP : nat -> bool
FnInP(n) ==
  is not yet specified;

end P
```

```
class Q

instance variables

possiblyR : [R];

operations

public
OpInQ : real * bool ==> nat
OpInQ(r,b) ==
  is not yet specified;

end Q
```

```
class R

instance variables

protected theOrderedQ : seq of Q;
private attribR : bool;

end R
```

Solutions for Chapter 11

Exercise 11.1

Some stakeholders for an enterprise architecture capable of supporting this policy are:

- the person responsible at the political level (usually a minister);
- the person/department responsible at the operational level;
- the traffic operator in a traffic centre;
- the road user;
- companies from the traffic engineering industry;
- companies from the software industry.

Solutions for Chapter 12

Exercise 12.1

```
per Put => #fin(Get) = #fin(Put)
```

Inductively, this is correct.

Exercise 12.2

```
per Put => data = nil and #active(Get) = 0;
per Get => data <> nil and #active(Put) = 0
```

Exercise 12.3

```
mutex(AcquireLock);
mutex(ReleaseLock);
mutex(AcquireLock, ReleaseLock, IsLocked);
```

Exercise 12.4

`#fin(ClientSend)` and `#fin(ServerListen)` must be equal, on the basis of `ClientSend` and `ServerListens` permission predicates. This will be one greater than `#fin(ServerSend)` (from `ClientSends` permission predicate). The answer is therefore:

```
per ServerSend => #fin(ClientSend) = #fin(ServerListen)
                  and
                  #fin(ServerListen) - 1 =
                  #fin(ServerSend);
```

Exercise 12.5

There are many ways of thinking about this. One way is to consider the communications round. After `ClientListen` completes, `ClientSend` should be enabled. From the `ClientSends` permission predicate it is therefore clear that when the operation `ClientListen` is activated, the number of times it has completed is one fewer than the number of times the other operations have completed. Because the number of times each of the other operations has completed is the same for each of them, taking an arbitrary operation yields

```
per ClientListen =>
    #fin(ServerSend) - 1 = #fin(ClientListen);
```

Exercise 12.6

```
class Buffer
instance variables
  data : [seq of char] := nil

operations

public Put : seq of char ==> ()
Put(newData) ==
  data := newData;

public Get : () ==> seq of char
Get() ==
  let oldData = data
```

```
   in
   ( data := nil;
     return oldData
   )

sync
   per Put => data = nil;
   per Get => data <> nil;
   mutex (Put, Get);
   mutex (Put);  -- only difference
```

Solutions for Chapter 13

Exercise 13.1

```
public int validateOp(int n)
   {
     try {
       interp.DynPreCheck(true);
       VDMFactory fact = app.GetVDMFactory();
       VDMSequence args = fact.MkSequence(client);
       VDMNumeric intValue = fact.MkNumeric(client, n);
       args.ImpAppend(intValue);
       VDMGeneric result = interp.Apply(client, "a.op",
                                        args);
       VDMNumeric number = VDMNumericHelper.narrow(result);
       return (int) number.GetValue();
     } catch (APIError e) {
       System.err.println("Unable to validate model");
       return -1;
     }
   }
```

Exercise 13.2

```
public void shutdown(){
     try {
       app.DestroyTag(client);
```

```
      app.Unregister(client);
   }
   catch (APIError e) {
      System.out.println(e.msg);
   }
}
```

Exercise 13.3

It is not possible to get the result of an operation call using the EvalCmd method; therefore it could not be used to obtain the server's response.

Exercise 13.4

```
private boolean checkResponse(VDMGeneric
                                genericResponse)
{
   try {
      Log("response is " + response.ToAscii());
      if (genericResponse.IsNil())
         return false;
      VDMRecord response =
              VDMRecordHelper.narrow(genericResponse);
      return response.GetTag().equals(
              "POP3Types'OkResponse");
   } catch (APIError e) {
      System.err.println(e.toString());
      return false;
   }
}
```

Exercise 13.5

1. This would just involve replacing the string ".ClientListen" with the string ".ClientWait" in the sendCommandResponse method.
2. This would require that the string "mk_POP3Types'" is replaced with "new" in the method makeCommand.

Exercise 13.6

Two situations in which this approximation is insufficient:

1. If when the Toolbox was started, another model was loaded and initialised, then the interpreter will contain the wrong model.
2. If a client was started but then killed between loading the VDM++ model and initializing the interpreter, then the interpreter will not contain an initalised model.

Solutions for Chapter 14

Exercise 14.1

Because the name `equals` is already used (by `java.lang.Object`), the method is renamed `vdm_equals` in the generated code.

Exercise 14.2

```
public boolean equals(Object obj)
{
  if (obj instanceof A)
    return vdm_equals((A) obj);
  else
    return false;
}
```

References

[APIMan] The VDM Tool Group. *VDM Toolbox API*. Technical Report, IFAD, October 2000. Visit the book's Web site for further information.

[Beizer90] Boris Beizer. *Software Testing Techniques, second edition*. Van Nostrand Reinhold, 1990.

[Bicarregui&94] Juan Bicarregui, John Fitzgerald, Peter Lindsay, Richard Moore and Brian Ritchie. *Proof in VDM: A Practitioner's Guide. FACIT*, Springer-Verlag, 1994.

[Bjørner&82] D. Bjørner and C.B. Jones, editors. *Formal Specification and Software Development*. Prentice-Hall International, 1982.

[Boehm81] Barry W. Boehm. *Software Engineering Economics*. Prentice-Hall, Inc., 1981.

[Booch&99] Grady Booch, Ivar Jacobson and Jim Rumbaugh. *The Unified Modelling Language User Guide*. Addison-Wesley, 1999.

[CGManJavaPP] The VDM Tool Group. *The VDM++ to Java Code Generator*. Technical Report, IFAD, October 2000. Visit the book's Web site for further information.

[Cockburn00] Alistair Cockburn. *Writing Effective Use Cases*. Addison Wesley, 2000.

[Fitzgerald&98] John Fitzgerald and Peter Gorm Larsen. *Modelling Systems – Practical Tools and Techniques in Software Development*. Cambridge University Press, 1998.

[Fowler99] Martin Fowler. *UML Distilled: A Brief Guide to the Standard Object Modeling Language*. Addison Wesley Longman, 1999.

[Gamma&95]   Erich Gamma, Richard Helm, Ralph Johnson and John Vlissides. *Design Patterns – Elements of Reusable Object-Oriented Software*. Addison Wesley, Pearson Education, 1995.

[Gosling&00]   James Gosling, Bill Joy, Guy Steele and Gilad Bracha. *The Java Language Specification*. Addison Wesley, 2000.

[Harris95]   Robert Harris. *ENIGMA*. Arrow Books Limited, 1995.

[Hodges92]   Andrew Hodges. *Alan Turing: The Enigma*. Vintage, Random House, 1992.

[IEEE1012-1998]   *Standard for Software Verification and Validation – Description*. IEEE, 1998.

[IEEE1471-2000]   *Recommended Practice for Architectural Description for Software-Intensive Systems 2000*. IEEE, 2000.

[ISO8402]   Quality Management and Quality Assurance – Vocabulary. 1994.

[ISOVDM96]   P.G. Larsen, B.S. Hansen, H. Brunn, N. Plat, H. Toetenel, D.J. Andrews, J. Dawes, G. Parkin, et al. Information Technology – Programming Languages, Their Environments and System Software Interfaces – Vienna Development Method – Specification Language – Part 1: Base Language. December 1996.

[Jones80]   Cliff B. Jones. *Software Development: A Rigorous Approach*. Prentice-Hall International, 1980.

[Jones90]   Cliff B. Jones. *Systematic Software Development Using VDM, second edition*. Prentice Hall, 1990.

[Jones99]   C.B. Jones. Scientific Decisions which Characterise VDM. In Jeannette M. Wing, Jim Woodcock and Jim Davies, editors, *FM'99 - Formal Methods, Volume I*, pages 28–47, Lecture Notes in Computer Science 1708, Springer-Verlag. 1999.

[Kruchten95]   Philippe Kruchten. Architectural Blueprints – The 4+1 View Model of Software Architecture. *IEEE Software*, 12(6):42–50, November 1995.

[LangManPP]   The VDM Tool Group. *The IFAD VDM++ Language*. Technical Report, IFAD, April 2001. Visit the book's Web site for further information.

[Pressman04]   Roger S. Pressman. *Software Engineering: A Practitioner's Approach , sixth edition*. McGraw-Hill, 2004.

[RFC1939]   J. Myers and M. Rose. *Post Office Protocol - Version 3*. Technical Report, 1996. http://rfc.sunsite.dk/rfc/rfc1939.html.

[Rosenberg&99]   Doug Rosenberg and Kendall Scott. *Use Case Driven Object Modeling with UML: A Practical Approach. Object Technology Series*, Addison Wesley, 1999.

[Shaw&96]   Mary Shaw and David Garlan. *Software Architecture: Perspectives on an Emerging Discipline*. Prentice Hall, 1996.

[Singh99]   Simon Singh. *The Code Book*. Fourth Estate Limited, 1999.

[Sommerville04]   Ian Sommerville. *Software Engineering, seventh edition*. Addison-Wesley, 2004.

[Unicode]   Unicode Editorial Committee. *The Unicode Standard Version 3.2*. Technical Report, 2002. http://www.unicode.org/unicode/reports/tr28/.

[UserManPP]   The VDM Tool Group. *VDM++ Toolbox User Manual*. Technical Report, IFAD, October 2000. Visit the book's Web site for further information.

[Zachmann87]   John Zachmann. A Framework for Information Systems Architecture. *IBM Systems Journal*, 26(3), 1987. See also www.zifa.com.

List of Acronyms

API	application programming interface
CBD	component-based development
CORBA	Common Object Request Broker Architecture
COCOMO	constructive cost model
COTS	commercial off the shelf
CPR	central population register
CSL	control speed limitation
CSLaM	control speed limitation and monitoring
CWS	congestion warning system
DB	database
DSI	delivered source instruction
EJB	Enterprise Java Beans
EPR	electronic patient record
FME	Formal Methods Europe
GOFO	good old-fashioned office
GUI	graphical user interface
HTTP	hypertext transfer protocol
IDL	interface description language
IEEE	Institute of Electrical and Electronics Engineers
IT	information technology
JFITS	Japan Future Information Technology & Systems Co., Ltd.
JSP	Java Server pages
LAN	local area network
OCL	object constraint language
OMG	Object Management Group
OO	object oriented
ORB	object request broker
POP	Post Office Protocol
RUP	Rational Unified Process
SAN	storage area network
SA/SD	structured analysis/structured design
SEI	Software Engineering Institute
UML	Unified Modeling Language

VDL   Vienna Development Language
VDM Vienna Development Method
VM   virtual machine
VPN   virtual private network

Subject Index

Definitions Index